GLOBAL SUSTAINABILITY INSIDE AND OUTSIDE THE TERRITORY

Proceedings of the 1st International Workshop
Benevento, Italy   14 February 2014

Editors

Carmine Nardone *(Futuridea, Italy)*
Salvatore Rampone *(University of Sannio & Futuridea, Italy)*

World Scientific

NEW JERSEY · LONDON · SINGAPORE · BEIJING · SHANGHAI · HONG KONG · TAIPEI · CHENNAI

GLOBAL SUSTAINABILITY
INSIDE AND OUTSIDE
THE TERRITORY

Published by

World Scientific Publishing Co. Pte. Ltd.

5 Toh Tuck Link, Singapore 596224

USA office: 27 Warren Street, Suite 401-402, Hackensack, NJ 07601

UK office: 57 Shelton Street, Covent Garden, London WC2H 9HE

Library of Congress Cataloging-in-Publication Data
Global Sustainability Inside and Outside the Territory (Conference) (1st : 2014 Benevento, Italy)
    Global sustainability inside and outside the territory : proceedings of the 1st international
workshop / Salvatore Rampone, University of Sannio, Italy, and Carmine Nardone, Futuridea, Italy
[editors].
        pages cm
    "This book contains the Proceedings of the Workshop 'Global Sustainability Inside and Outside
the Territory' held on February 14, 2014, in Benevento, Italy..."--Foreword.
    Includes bibliographical references and index.
    ISBN 978-9814651318 (hardbound : alk. paper) -- ISBN 9814651311 (hardbound : alk. paper)
    1. Sustainability--Congresses. 2. Sustainable development--Congresses. I. Rampone, Salvatore,
editor of compilation. II. Nardone, Carmine, editor of compilation. III. Title.
    HC79.E5G5971485 2015
    338.9'27--dc23
                                                                                    2014042892

British Library Cataloguing-in-Publication Data
A catalogue record for this book is available from the British Library.

Cover image: photo by Salvatore Purificato processed by Salvatore Rampone

Printed in Singapore

FOREWORD

This book contains the Proceedings of the Workshop 'Global Sustainability Inside and Outside the Territory' held on February 14, 2014, in Benevento, Italy, organized by Futuridea – Innovazione Utile e Sostenibile with the sponsorship and support of Global Sustainable Social and Energy Program – GSSEP Onlus, Department of Science and Technology – University of Sannio, CNR ISAFoM, ConSDABI – Consorzio per la Sperimentazione Divulgazione e Applicazione delle Biotecniche Innovative, Confederazione Italiana Agricoltori, Fondazione ITS BACT, Provincia di Benevento, Confindustria Benevento, Camera di Commercio Benevento, and Consorzio ASI Benevento.

The purpose was to getting together researcher and stakeholders interested in communicating, establishing collaboration, exchanging ideas in the global sustainability area, which calls for the widest possible interdisciplinarity.

The proceedings consist of 10 invited and contributed papers related to the broad range of aspects of sustainability in a global scenario including food safety, monitoring, soil mapping, health care, territorial intelligence, local food production, greenhouse gas emissions, renewable energy sources, integrated development, sustainability strategies, 'smart' bio-territories.

The workshop was largely a success and led to the signing of international agreements for the protection and enhancement of endangered species in the area of North Africa. We wish to thank all those who have contributed to organize and realize this event. Our appreciation is due also to all those Institutions and Companies that in different ways joined us in our job.

Salvatore Rampone
Scientific Director, Futuridea

Contents

Organizing – Scientific Committee:

Carmine Nardone (Futuridea – Innovazione Utile e Sostenibile), Carlo Sinatra (Global Sustainable Social and Energy Program – GSSEP Onlus), Salvatore Rampone (Department of Science and Technology, University of Sannio and Futuridea – Innovazione Utile e Sostenibile), Antonio Leone (CNR ISAFoM), Gianni D'Angelo (Department of Science and Technology, University of Sannio), Maria Luisa Varricchio (Futuridea – Innovazione Utile e Sostenibile), Hansruedi Schenk (Global Sustainable Social and Energy Program – GSSEP Onlus), Wanda Ternau (Global Sustainable Social and Energy Program – GSSEP Onlus), Massimo Squillante (Department of Law, Economics, Management and Quantitative Methods, University of Sannio)

The sponsorship and support of:

Futuridea – Innovazione Utile e Sostenibile
Global Sustainable Social and Energy Program – GSSEP Onlus
Department of Science and Technology, University of Sannio
CNR ISAFoM
ConSDABI – Consorzio per la Sperimentazione, Divulgazione e Applicazione delle Biotecniche Innovative
Confederazione Italiana Agricoltori
Fondazione ITS BACT
Provincia di Benevento
Confindustria Benevento
Camera di Commercio Benevento
Consorzio ASI Benevento

are gratefully acknowledged

A GREATER SUSTAINABILITY IS POSSIBLE

Carmine Nardone

President of FUTURIDEA, Georgofili Ordinary Academic
carmine.nardone@gmail.com

Maria Luisa Varricchio
Research Staff Member FUTURIDEA
varricchio@unina.it

Abstract

Encouraging a greater sustainability at both local and global level is possible. The conditions making this aim possible concern different phenomena (population growth, desertification and food production) that must be taken into account with their complexity and interconnections in order to cope with present emergencies by adopting practical solutions. New technological solutions are necessary and indispensable for an eco-conversion of productive systems. At the same time a structural upgrading of food production is needed. Greater sustainability cannot be achieved without re-considering the 'cycle', the extraordinary mechanism whereby organic matter is destroyed in breathing and regenerated in photosynthesis. Questions worth being discussed and investigated range from renewable energy for the new agriculture to the reduction of fossil fuel consumption for the production of food energy, low CO_2 emissions, new forms of protection of agricultural soil and water resources, in contrast to the ongoing processes of desertification and erosion of unrenewable natural resources (plant and animal biodiversity rural landscapes, etc.). Unsustainability of land grabbing should give cause for reflection too. On the other hand, the global sustainability project should involve issues such as the risks and opportunities related to the activity of food production within the urban and metropolitan areas conurbations or in arid or semi-arid areas, also the questions related to desalination, efficient use of water resources, role of the new eco greenhouses deserve attention.

1 Introduction

The report "Our Common Future" of the World Commission on Environment and Development, also known as the Brundtland Report, issued in 1987 within the United Nations Environment Programme, provides a definition of 'sustainable development' resulting from an extended debate which had taken in the previous years. Sustainable development is meant as a development capable to meet the needs of the present without compromising the capability of future generations to meet their own needs. This Report outlines the importance of a new relation between economic activity and natural resources, which sets limits to income production at a level [1,2] that does not damage non-reproducible natural resources. The fundamental theme of intergenerational rights is dealt with for the first time in an official report institutional report in a clear way. This definition will be conveniently referred to in numerous subsequent institutional acts [3–5]. This assumption, of extraordinary importance, rises two questions to mankind: the first regards the complexity (institutional, economic, social and environmental) related the aim of pursuing sustainable development, local and global, as a systemic action; the second breaks the path for research and thinking on the theoretical and practical conditions for greater sustainability.

The first point suggests not to separate the issue of sustainable development from the issue of the global structural crisis that economy is living nowadays. In the present situation the economic system is providing future generations with three unprecedented records: the highest unemployment rates in history, a terrible record in malnutrition, the highest debt load of both Western countries and developing countries. These three phenomena deriving from the global crisis never reached similar levels and, above all, never interacted so strongly [6].

The crisis, therefore, has been and will be devastating for the years to come in all fields: economic, social and environmental. And it will have long-term effects.

On the one hand, there will be an increase both in unemployment and in the number of people at risk of poverty in developed countries on a global scale and, on the other, the growth of public debt involving a major reduction of resources for social policies and environmental policies in place.

The vast majority of choices that have been taken in recent years caused a series of appalling situations such as the dramatic poverty, destruction of non-reproducible natural resources, water crisis and energy and a huge growth in the illegal economy.

The public debt (according to the Bank for International Settlements (BIS) – 100 trillion, which is equivalent to almost one and a half times the wealth created on Earth in a single year) – represents not only an opportunity for the leaders of illegal cartel economy but also for recycling funds and, in general, for illicit proceeds and flows from tax evasion and drug trafficking with the apparent complicity of the world of global finance.

The financial crisis of recent years has made everything more dual, broadening the gap between the rich and the poor in the world, with the contradiction that competing "capitalisms" increase the monopolistic voracity and extend their increasingly all-encompassing control over innovations and on the distribution of goods and services.

Within this general frame the basic contradiction between multinational and transnational development (exclusively oriented towards profit accumulation with no social responsibility) and, on the other hand, then nation states increasingly in trouble, with less power and fewer resources [6,7] emerges.

The present state of the global economic and social conditions might justify the suggestions of the negative thinking of those scholars who consider the environmental catastrophe as inevitable for humanity. James Lovelock [8] is one of these. The English chemist and biologist, father of the "Gaia hypothesis", also known as "Gaia theory" or "Gaia principle", deems unavoidable to stop the process of global warming. The situation is so serious that the efforts undertaken for the production of alternative energy appear useless. According to him: "It is like rearranging the deck chairs on a sinking ship".

In his opinion, for the future we should aim in particular to the desalination in order to face the severe water crisis that risks to become more dramatic in the coming years with the advance of desertification and, on the other hand, we should try to produce food in a more sustainable way. These goals are at the core together with the issue of reducing CO_2 emissions for a more sustainable development both locally and globally.

Today we have reached a record level of CO_2 in the atmosphere: 400 parts per millions [9]. These data are even more worrying given the reserves of fossil fuels, 2,795 trillion tons of CO_2 on one side and emission limits to avoid the rising of the temperature of the planet on the other. The expectations of the lending banks financing the 100 largest oil companies in the period 2000 to 2012 (with about 7000 billion dollars) are not compatible with emission limits provided for by the Kyoto Protocol.

Without a profound structural change, we are faced with an agonizing choice between a 'carbonate bubble' and a new 'financial bubble'. Never before has been so decisive for the future of humanity to proceed with force to those innovations and technological solutions capable of supporting a concrete orientation towards greater sustainability. The truth is that, luckily, there are solutions and they are becoming increasingly available. The point is that the development paths of the last century (capitalism and centrally planned economies) were structurally oriented towards un-sustainability with the destruction and the depletion of natural resources that cannot be reproduced.

For these reasons, 'sustainability' on the one hand is a real new development model and on the other it finds strong barriers and antagonisms in the wake of the old cultural models. The claim that more sustainability is possible comes mainly from the consolidation of an increasingly technological offer making more compatible eco reconversion and competitiveness. This, surely, is a phenomenon that tends to consolidate and affirm only that a sustainability-oriented path is technically possible. Are the choices of local and global elites to remove those political barriers that today make it difficult the emergence of new local and global conditions, and especially to provide the indispensable new world economic rules. Encouraging a greater sustainability is possible also because if not all people and bodies that are currently fighting for a different development both locally and globally would be deprived of their strength. Of course, this positive orientation cannot be confused with an unreasonable optimism, indeed must increase awareness of both the urgency and the powers and antagonistic barriers to this path. It is important to raise the awareness that we have no time to simply observe and wait but it is necessary to disseminate what has been already made available and useful by the research. Sustainable innovations must be promoted in order to overturn the dominant paradigm, based on an indeterminate and limitless growth. Moreover, nowadays public and private bodies do not give the word sustainability the right importance. On the contrary, they use it as a label just for marketing reasons that have nothing to do with a real change. Urgent and desirable reflection is also necessary within those forces who believe in sustainability even if they don't agree on strategies to be adopted. There are cultural sectors influenced by some environmental movements who think they can achieve sustainability by going back to the past as if the "old" was synonym to sustainability. These obscurantist positions deny, in fact, the essential value of biological and cultural evolution. "Evolution" means, first of all, differentiation and transformation, i.e. increase in variety of types available. Typically (but not always) this corresponds to an increase of complexity. Finally, it means development of skills and new interaction with the environment [10].

The path to a new system is ineluctable. UN reports on the expected growth of the World population speak clearly. In the coming decades, the planet will have to support a population increase by about 5 billion people. The growth will concentrate in urban centres whose the population will constitute about 80% of the total global population [11][1]. These data become quite alarming if evaluated together with the reports of the UNCCD (United Nations – Convention to Combat Desertification) [12], which warns against the risk of depletion of arable land, the rising pollution and irrational use of water resources. In Southern Italy, the impoverishment of the soils approximately affects 18% of the UAA (utilized agricultural area). According to Franco Miglietta, a researcher at the Institute of Biometeorology of the National Research Council (Ibimet-Cnr) – The Italian agricultural soils traditionally contained 130 tons of carbon per hectare, today almost half. This is the most obvious result of the missed closing of the local cycles [13,14]. The decline in the availability of carbon, of course, is not a phenomenon only Italian, but it concerns, with more or less intensity, all ground in particular of Western countries.

In this paper we try to provide useful elements for an integrated and functional idea of what complexity really means. On the one hand, we want to put forward the thesis that it is impossible to separate the issues related to earth, energy, rural landscapes, social equity and new human and social rights. On the other hand we give a great importance to those useful and sustainable technologies that can address positively the three major emergencies that the world must deal with: soil, energy and food.

2 Land and Sustainability

The prevailing development models have been mainly based on the indiscriminate extraction of organic matter from the soil with the effect of creating its progressive impoverishment it. Among the many emergencies, the relationship between soil and sustainability becomes a top priority.

Today, more than ever, a structural conversion is needed involving a comprehensive process oriented to close the cycles (water, oxygen, carbon, nitrogen and phosphorus) that is to act in such a way that all natural transformations, alimented by solar energy, will force the matter involved in the cycle to be continuously reused. In other words everything should become raw material for other natural cycles.

After the amazing message of the American biologist Barry Commoner, author of "Four laws of ecology" [14], it became impossible not to take into

---

[1] The report State of the World, designed by Lestern Brown, stands in the vast international literature because it offers a systemic vision and dynamics, starting from the year of first publication in 1984, until the publication of the last report in 2013 from the significant title: Sustainability Is Still Possible?

account idea of "closed cycle" in nature – perhaps altered by man, who has upset the ecological balance by preventing the closure – and to fully assume the complexity and responsibility for the sustainability of the "life" phenomenon. Agriculture is made of cycles that opened and closed on a local basis. With the current models, it is no more so; devastating imbalances are the result[2]. The land has been exploited and degraded both in market economies and planned economies for not having put limits to the consumption of soil by the wild urbanisation and for not imposing proper protocols use for agricultural exploitation.

FAO data of 2014 [15] give an idea of the extent of the phenomenon:

– Currently the agrarian world usable area is approximately 4.4 billion hectares;
– 1/4 of the land surface of the planet is threatened by the phenomenon;
– 3/4 of the arid lands in North America and Africa, are at high risk of desertification (and this clearly demonstrates that the phenomenon is not just a problem for the African areas, but also for parts of North America and, in some cases, Canada and many areas of Europe;
– 900 million human lives are threatened by desertification in Africa;
– 20% of irrigated agricultural land, out of a total of 250 million hectares worldwide, is affected by salinization, a real antechamber of desertification;
– On average 10 million hectares of forests each year are destroyed due to fire or to change of land use. The planet, on the one hand, is called upon to meet the needs of food for an increasing number of people (soon as much as 7.5 billion) on the other hand, it is subject to a devastating global degradation of farmland which is not limited only to some areas of the world, but which affects, with different intensity, the entire surface of the planet. The population grows and the fertile land decreases. Every year the world population grows more than the Italian population and at the same time the fertile land decreases as the Italian agricultural surface (12.4 million acres). Thus, drastically decreases the availability of arable land per capita in the world. In 1900, considering the world's population (1,522,000,000) and the agricultural land available (about 5.000.000.000 hectares) the arable land availability per inhabitant stood at about 3.28

---

2 The Milan Expo 2015, "feeding the planet, energy for life" represents an outstanding opportunity for starting a new global and local horizon, able to stimulate a real eco-revolution of how to grow food.

hectares. In 2005 the availability decreased dramatically, reaching about 0.68 hectares of arable land per capita [16][3]. In 2014 the availability was further reduced to approximately 0.60 ha. In 2050 it will drop significantly, due to population growth and the decrease of agricultural surface used (estimated 0.46 hectares). On May 23 2007 (according to other sources in 2009) the urban populations have outstripped the rural ones. In 1900 the urban population was 13%, and in 2007 became 51%. The data provided by the UN predict that in 2050 the percentage of urban populations will be 75%.

The phenomenon also affects the Mediterranean basin where an important process of desertification is caused by the inevitable pressure on natural ecosystems resulting from population explosion. In the Mediterranean countries, in fact, population has grown from 90 million inhabitants (early last century) to the current 300 million. According to the most optimistic forecasts, it is expected to increase to nearly 850 million by 2050.

The phenomenon of desertification is expected to worsen as a result of ongoing climate change. The most authoritative climatology institutions, such as the Met Office Hadley Centre (UK) and the Potsdam Institute for Climate Impact Research (Germany), on the basis of a "business as usual" scenario, foresee for the Mediterranean area a temperature increase ranging from 2–4°C and a reduction in precipitation of about 1 millimetre per day.

It is clear then, from this general framework, that the interweaving of different phenomena (population growth – desertification – food production) leads to a major complexity that has to be taken into account in order to face the problem with concrete actions.

The simultaneous population growth and desertification of soils is the basis of a gigantic related phenomenon known as Land Grabbing [17,18] that is the land grab by corporations and investors.

These are purchases by – or long run concessions to – corporations and investment funds or foreign government of the right to exploit arable land. Corrupt local bureaucracies mostly at the expense of local populations facilitate the phenomenon. The effects of this wild land rush, are culpably underestimated by local government authorities. These gigantic purchases involve more often the uprooting of local communities from the land that they have occupied for thousands years. The new poverty produced increasingly pushes the expelled

---

[3] FAO data compiled and published by the technical staff of Futuridea in: Useful and sustainable Innovations, the Repertoire of Useful Innovation and good practice. Italy 2014.

rural communities to find shelters in urban areas increasingly degraded for lack of hygienic and other services [19][4].

For that minority of farm workers who remain in the new settlements the effect is inhuman exploitation. In many cases it is difficult to achieve even a subsistence wage. It is a deadly attack peasants' affection to the land that has historically been a real act of love for nature. For millennia, man has ensured proper ecobalance of territories, as if our ancestors were anxious to experience the importance of "biocapacity" ecosystems.

The history of agriculture was made by those who for more than 11,000 years have built the future of mankind producing food and devoting their life to agricultural work. The landscape and the agricultural sciences have taken advantage of the humble and tenacious relationship of women and men with the Earth. Despite the love, the land was usually a chimera for the peasants. The impact of the land rush not only produces social effects on site but feeds unfair competitions of global character and deconstructs irreversibly rural landscapes and valuable natural resources, millennial developments related especially of plant and animal biodiversity [20]. The protection of land and of its fertility on a global scale becomes crucial to humanity and a primary local common concern of humanity. This is the only way to ensure a future of well-being and food for future generations. It is time to give practical effects to the intentions by using on a large scale also those new technologies that make possible new soil maps (satellite data, real time sensors, Vis-NIR spectroscopy) in order to monitor the state of the soil and to use the information system as the basis for a "precision agriculture".

3 Food

In this context of the wild ride to the land FAO points out that: "The increase in world population and the increasing demand for food pose great challenges to agriculture."

In coming years we will have to produce more food using fewer natural resources and responding to climate change. "UN agency", in fact, estimated that world food production must increase by 60% by 2050, mostly on land already cultivated. In addition to the question of the agrarian soil it is necessary to face the drastic the current distortions in the present production and consumption of food in order to ensure the food of the future.

---

[4] It is a scientific work on the social effects of globalization in the food sector which can be regarded as 'ringleaders' of the subsequent literature.

One of the symbols of the complexity and the changing relations between agricultural and food production and soil is without any doubt the waste of food in the world.

A huge amount of organic substance taken in vain that returns to earth and take the road of landfills. The latest figures speak of about two billion tons. Of course, not all waste can be recuperated: the percentage can vary depending on the nature of the products.

New technical protocols for the storage of food products, such as the use of so-called systems of "modified atmospheres" [21] of great importance for the small and medium enterprises, more and more oppressed by monopoly of the multinational holdings of agri-food production and transformation. At the same time, it is necessary to enhance the alternative use (quality composting or other innovative forms of recycling) of the waste non-reusable for food purposes. A second essential element for a sustainable future is the link between food and biodiversity. The relationship between biodiversity, food and quality of rural landscape are absolutely evident. An area rich in biodiversity is also a territory with more quality of the landscape. Genetic erosion has also meant the loss of landscapes.

According to FAO [15], 120 species provide 90 percent of foods. Only 12 species and 5 species of animals provide 70 per cent of the food. Only 4 plant species (rice, maize, potato and wheat) provide more than 50 per cent and three species of animals (pigs, cattle and chickens) provide more than half. Food merely standardizes the well-being and inhibits potential to customize the food models (nutraceuticals) [22].

Humanity has also called for a style food can increase expectations of well-being and life and, on the other hand, produce a more sustainable food, also means more environmental well-being. According to recent studies by FAO, excluding fishing, mankind consumes 56 billion animals a year. In Western countries, the consumption of meat exceeds 80 kg per year (global average 37). A Western citizen consumes 7.36 times the quantity of meat consumed by a citizen of Africa. The water necessary to produce the amount of meat is the highest ever (15,400 gallons per kg) without counting inputs of fertilizers and pesticides for the production of animal feed. A kilogram of legumes (beans, lentils, chickpeas, broad beans, etc.) on average consumes only about 200 litres. More vegetable protein and less animal protein, however to produce in a more environmentally virtuous eco-environment. The nature of processes that govern the modern world requires a cultural revolution: in the near future we should all look at every phenomenon as a manifestation of global dynamics.

The concept of "local" requires adequate redefinition in order not to refer to theoretical and practical economic and administrative models which might be limiting. The development of the local society is based on a project that requires the overcoming of the territory as a mere support of economic activities or as a soil-resource to be consumed within the idea of unlimited growth.

The question to which we are required to answer is: will we be able to provide food in sufficient quantities to meet the in a sustainable manner the demand of food and water, which tends to reach unprecedented levels such inedited in the history of mankind?

This question has sparked a debate increasingly fueled by positive solutions and technical contributions. It is processing that provide new insights, cognitive elements, new design solutions and ideas on key aspects of the theme of sustainable development and, in particular, of the food-energy nexus-population growth, capable of dealing with the consumption of the ozone layer, land erosion, genetic erosion, degradation of non-renewable resources, desertification, deforestation, water scarcity, poverty, unemployment.

The overall effort is essential to define a new carrying capacity of the planet, that is to say a complex system of variables: technologies, climate, environmental impact, population distribution etc.

The increasing unsustainability of urban concentrations, with air pollution that is more and more uncontrollable, requires choices that cannot be longer postponed in order to switch from emergency and short-term decisions to structural changes, globally and locally. Rethinking the precondition means to rethinking, without delay and with all the possible radicalism, the relationship between city and country, and grasp all the implications of a world that has as a realistic prospect the 50% of the population massed in new forms of urban gigantism and the rest of the population in rural areas increasingly desertificated and depleted.

The first crucial question, is therefore to review and foresee a way to produce food: to produce more food in urban centres and in a more sustainable way (hydroponic solar greenhouses, green centre of agricultural production and multi-functional networks of condominium gardens) and to use of land for the development of massive tree crops (able to give greater sustainability for their ability to break down CO_2) and legumes (nitrogen-fixing plants). In practice, it is a profound process of balancing urban and rural contexts thanks to less energy consumption for transportation by shortening the distance between the places of production and the places of consumption of food, All this can be achieved reducing energy consumption in agriculture, reducing the use of fertilizers and pesticides and using less energy from fossil fuels and more renewable energy at the farm level.

This requires also a strong commitment of modern architecture in buildings-greenhouse in urban areas of metropolises in the world, able to shorten the supply chain between food production and consumption, and to rationalize resource use preferably generated from renewable and non-polluting sources. So creativity can explode in vertical architecture suggesting forms that anticipate high-rise buildings capable of producing food for the inhabitants. The idea of building concrete skyscrapers to produce food or urban forests (from New York, Toronto, Milan, etc.) appears as a race to the "gigantism", a choice questionable in terms of sustainability. The redevelopment of many disused industrial sites would be much more sustainable. The growing complexity of agricultural food systems is of great interest the positive contribution of new genetic innovation techniques. This is the technique known as MAS (Marker Assisted Selection) that allows the genetic improvement of "optimizations", without introducing genetic fragments "strangers" in the genome of individual plant species [23]. The potential of this technique are important: first, because it allows you to 'customize' genetic improvement to local environments and also allows to select a variety of different plant species that can grow in an integrated manner or hold together a greater capacity for resistance to parasitic attacks with increased tolerance to salinity and with improved nutritional quality of foods. It is a practical alternative, non-ideological. In addition to ethical aspects there are economic benefits for agricultural producers, because these cultivars are not patentable and, as such, exempt from royalties conveying machines biotechnological incomes [24,25].

4   Rural Landscape

Encouraging food sustainably is possible only in areas where natural resources are constantly preserved with the consequent retention of biocapacity of the eco-system. The rural landscape must be seen according to the new functionality, i.e. the vital frame of the agro mosaic able to give back what the agricultural production takes. The quality of vegetational areas and non-productive rural areas oriented to biodiversity (edgings, soil protection by landslides) becomes essential. According to the most accredited recent theoretical references, in some way is not only representative of the structural situation. Subjectivity and the local communities produce a "sense effect" of the landscape. One can say, in other words, that the landscape includes both the reality and the representation[5] [26]. The "perception" is not unique, of course, in can vary even within the same local communities but can vary also among different subjects in relation to numerous variables (sex, age, social status, etc.).

Globalization and social-networks introduce new virtual perceptions towards new landscapes, on the other hand weaken the classical local perception "identity". The effects of social networks are still to be explored and analysed. So, now, it is only possible to highlight some dynamics:

a)   exponential growth in the net of narratives of landscapes on a global scale (photos, movies, testimonials etc.) which feeds the knowledge and curiosity, albeit in a virtual, for an unlimited number of landscapes; and

b)   virtual relations between migrant populations and the lands of origin, causing renewed emotions and different perception of the landscape of origin;

c)   definitions of new apps that can enhance the knowledge of the originality of the landscape to facilitate the appreciation and interaction with the "excellence" landscape for tourism and study

Perception and love for landscapes are vital for the future energy landscape in particular of rural areas[6] [27]. According to Hillman, one must repeat, about territories and landscapes, the idea of 'soul', recovering from the Greek culture the ancient concept of a nature which absorbs the thoughts and traditions of men that inhabits it from centuries or millennia. With this conviction, we talk about 'soul places'. It is a call to awaken from the "anaesthesia" and inability to try sensations that envelops our culture, to rediscover the "animistic" and therefore Pagan, concept according to which everything is alive, everything speaks to us. It is an act of faith in the beauty that one can return a sense, landscape architecture, cities, and to our own lives. In this respect our landscapes show, despite the obvious economic and social difficulties of 'vitality' unsuspected energies of nature and culture. The definition of "rural" is the result, then, a

---

[5] It is a research decidedly innovative promoted by the Rural Network in nine European countries (lead Italy) on the perception of young people in rural areas in between the ages of 17 and 20 years. The research shows a good level of satisfaction for rural areas and the agricultural work.

[6] In this book, Hillman speaks of the soul of places – and of the sense of beauty, and the need to preserve it – with the architect Carlo Truppi, in a dialog was born in a special place, Syracuse, on the occasion of a symposium on recovery of the island of Ortigia. It is a dialog that winds through a land border, and on different paths and twisted the psychologist and the architect must be in search of ideas and meanings that go beyond the boundaries between the disciplines.

conceptual evolution meeting that covered the landscape and "rurality" as well as the innovation introduced by the EU Convention (2000) on the landscape and on the "perception" of it. In view of the complex new conceptual framework one can say that a possible definition of 'rural' landscape is the following: ' ... the rural landscape is a spatial agro-ecosystem inclusive or integrated with the natural environment (soil, water, climate, landscape, natural resources, plant and animal biodiversity, biocapacity etc.) from the cultural landscape (painting, photography, poetry, prose, music, etc.) and anthropogenic action historically carried out by men in rural areas (production systems, work, technology, rural architecture, etc.) and 'subjective' perception of men and women. A propulsive thrust to the growth of bio-territories intelligent and the true dam against the invasion of wild models of overbuilding and massification against the destruction of rural areas.'

5 Reduction of the Consumption of Fossil Fuels

These objectives are closely related and can be achieved through the use of alternative energy sources instead of fossil fuels, the so-called Renewable Energy Sources (RES), being able to effectively meet the energy needs for productive activities. The adoption of technologies that exploit RES breaks down pollutant emissions even at the local level, thereby increasing the quality of the final agricultural product. In addition, the use of these technologies allows production companies to save costs of energy supply.

The choice of a more sustainable agriculture is inevitable to give food to the existing population and to the other five billion people that will increase the global demography in the coming decades.

Greater sustainability cannot be achieved without reconsidering the 'cycle', that is to say the extraordinary mechanism of organic matter that is destroyed in breathing and regenerated in photosynthesis.

After the charming message of the American biologist Barry Commoner, author of the "four laws of ecology", it could not help but reflect on the idea of "closed loop" in nature, perhaps altered by man, which has troubled the 'ecological balance' preventing closure, to assume fully the complexity and responsibility of the sustainability of the phenomenon of "life". Agriculture is made up of cycles that opened and closed on a local basis. With the current models this is no longer possible; devastating imbalances have been the consequences.

The production of food, therefore, will not be sustainable given the increase in population and urbanization, the impoverishment of the agricultural area exposed more and more to the phenomenon of desertification and generalized loss of organic matter, increasing the use of resources (water, energy, fuels) in areas with water shortages and pollution

One of the ways to rethink the overall condition of the concrete way of producing food is to produce food even in urban systems through agro-housing, urban regeneration, eco-greenhouses etc.

This requirement involves modern architecture in proposing greenhouses buildings in the urban centers of cities around the world, capable of shortening the chain between production of food and its consumption, and to rationalize the use of resources, preferably generated from renewable and non-polluting one.

So the creativity explodes in supporting vertical forms of architecture, with the provision of high-rise buildings able to produce food for the inhabitants. The idea to build skyscrapers of cement to produce food or urban forests (from New York, Toronto, Milan etc.), seems like a race to the 'gigantism' that is questionable in terms of sustainability. It would be much more sustainable the conversion of many brownfield sites in real biofactories. The 'Expo 2015, "Feeding the Planet, Energy for Life"[7] promises an extraordinary opportunity for the launching of a new horizon, both global and local, able to initiate a true eco-revolution of the way of producing food.'

6 Energy Autonomy

The objectives that lead to sustainable development are closely related, and can be accessed through the use of energy sources alternative to fossil fuels — the so-called renewable energy sources (RES), managing to meet energy requirements effectively for productive activities. The adoption of technologies that use renewable energy emissions also reduce locally, thus increasing the quality of the final product. In addition, the use of these technologies enables producers to save on energy costs. The crucial point is how to use the same renewable sources the eco-friendly way. The dissemination of agro-energy in

---

[7] Very significant innovative orientation for the future is without any doubt The Israel Pavilion of Expo Milan 2015 previewed to the press by the Commissioner-General Elazar Cohen Monday, 2 December at Villa Madama, in the context of the fourth Israeli-Italian Intergovernmental Summit in the presence of Prime Minister and Netanyahu. Designed by architect David Knafo, will extend into the area of 2,370 square meters over an area bordering the Italy Pavilion, will be coated with a vertical garden grown made in accordance with the rules of bioarchitecture with 100% recycled materials.

recent years had an exploit throughout Europe and the world with many contradictions. Vast fields with large photovoltaic structures have occupied agricultural soil and giant wind farms have attacked and devastated landscapes of rare beauty.

The agricultural lands were used for the installation of plants mainly medium/large to produce energy 'not intended' for agriculture that, in a more sustainable vision, could be a vector of stimulus to the use, perhaps, of new agricultural machinery electric zero emission.

Rather, the photovoltaic systems should be "integrated", under each profile, with the plants cropping systems, and not see them as "separate structure", but as a coordinated set of activities in which the production process rationalizing the use of resources (electricity, water, etc.) self-produced on-site (photovoltaic, small wind turbines, water treatment) to allow the natural growing out of context, or improve the quality of native crops.

It dates back to many years ago the debut of technologies for the cultivation of plants in different places from the natural environment. Case studies of artificial lighting of the plant have been carried on since the sixties in order to reconstruct the best environmental conditions for productive efficiency.

The greenhouses became the instrument able to grow crops in locations with environmental characteristics unfavorable to their development or adverse seasonal periods. At the same time researches and scientific studies on the various aspects of the crops so-called "soilless", "hydroponic" and "aeroponic" have developed. In these cases we do not use soil to use environments entirely created for plant production.

The most striking figure is, however, that now, even in industrialized countries, the energy used comes from fossil fuels (oil, natural gas, etc.) almost exclusively. This involves not only CO_2 emissions, but also less food quality and much higher production costs.

For this, a reasonable proposal for a real "sustainable breakthrough" is to undertake a radical technological reconversion of production in "greenhouse", directing the whole system towards the virtuous assembly of new technologies (solar, wind, water recovery storm water, wastewater treatment, LED lighting, electric vehicles, etc.).

New enterprises to vertical farming, the restructuring of the slums, the ecological reconversion of abandoned industrial sites, would lead to increase, along with the dissemination of agro-housing, a new possibility for urban systems to break down pollutants or contain devastating dynamics. To produce food with all the aforementioned elements would reduce the use of pesticides, herbicides, fertilizers and to drastically reduce the use of fuel for farm machinery. The same water reuse for agriculture would save a lot of water or groundwater.

7   Actions for Sustainability

In short, the questions that need answers and urgent solutions, i.e. priorities for a realistic path of eco-change of agriculture are the following:

– To allow food production within urban agglomerations.

Vertical farming, agro-housing urban gardens, are spreading quickly and is definitely desirable increased naturalization of urban systems. The crucial point is how to characterize these new processes in a context of 'total' sustainable growth (water resources, energy, recycling). The idea of producing food in new forms of urban gigantism (agricultural "skyscrapers", etc.) are definitely not desirable for the negative environmental impact. On the contrary, the "urban agriculture", as mentioned above, might be a way for the redevelopment of disused industrial plants (recovery of municipal waste) avoiding unsustainable gigantism. In other words, it is desirable for a modern agro housing as long as they strictly oriented new models of sustainability while avoiding a chaotic no rules

– Develop and promote food production in arid zones.

The integration of modern technology (active vegetation, eco greenhouses, desalination, water recycling, etc.) makes possible agricultural production in arid lands [28]. In particular, the production of energy from renewable sources, the rationalization of the use of resource and new cropping patterns may allow the production of food in arid areas where traditional agriculture would not have chance of success.

– Face the phosphorus emergency.

If the development-environment relationship requires a radical turning point. It is necessary to review the scale of emergencies. Rightly, some important scholars have emphasized the real emergence of the phosphorus cycle [29,30][8]. It is surprising how this dramatic emergency, which already produces profound effects on systems agro-food, remains a matter confined to academics and environmentalists, without arousing alarm and appropriate commitments on the

---

[8] Among the numerous studies, indicate that of David A. Vaccari and Giorgio Nebbia (professor emeritus of Exhibition at the University of Bari) on the crisis of phosphates

part of institutional, political and professional elites and trade unions. The crisis has already produced the raising of prices of fertilizers and restrictive policies of some countries (i.e. China), with a catastrophic perspective if not in the very short period, certainly within a century. The paradox is the growing scarcity and the contemporary phosphorus waste. Deep and structural modification of the way of agriculture have led to replace a capillary network able to save this precious element with production centers that do not allow to retain the mineral in the soil, which, instead, is disposed of through the waterways and lost at sea, causing many side effects on water quality and on the overall life of the ecosystem. The signs of the crisis are beginning to be heard for the proposed duties from China for export of phosphorus; the Asian country has imposed as a matter of priority its domestic consumption in anticipation of future shortages. It has factories all around the world, from Brazil to India, to supply difficulties. The United States are transforming slowly from exporters in phosphorus importers. Even Morocco reserves are not inexhaustible. An emergency, therefore, strong but silent, far from public opinion and, therefore, without credible investigations in progress in the study of solutions [6].

    – Sustainable intensification.
To promote a complex, multidisciplinary and interdisciplinary action capable of integrating new innovative technologies to increase productivity and, at the same time, exploit factors of production that are most threatened, i.e. water and soil in a sustainable manner the.

    – Promote the development of indigenous productions.
The increased domestic and international demand for high quality traditional Mediterranean products provide an important opportunity to increase the level of profitability of agriculture in Mediterranean countries and make it more globally competitive. This type of technology promotes traceability, which guarantees quality and safety to the consumer, with great impact on all the activities related to the production. Attention will also be drawn to the communicative aspect of the offered product, creating a marketing and providing transparent traceability and anti-counterfeiting through the use of digital technology.

    – Improving water use efficiency. Reusing waste of crops.
The adoption of techniques and irrigation methods that improve the efficiency of water use (water use efficiency) can contribute to the improvement of primary production, through the introduction of new species and a significant reduction

in water consumption, with positive consequences both in economic and environmental terms (sustainable irrigation).

– Reusing waste of crops.
The waste products of the crops can be used as organic fertilizers (compost on farm), with positive impact on soil quality.

– Improve the energy balance.
Through the reduction of the fossil fuel energy/food energy (currently up to 1/10 in the U.S.) as well as by reducing the total energy consumption of fossil.

– Safeguards rural landscapes.
– New rurality: dissemination of rural bio-architecture, rural building self-sufficient energetically; integration between rural productive areas (biodiversity of crops) and non-productive areas (biodiversity of wild species).

8   Conclusions

The potential for a more sustainable world does exist, but we don't have to underestimate the obstacles and powers against it. The crucial issue at a local and global level is the need for a new quality of development. However, the dominant multinational economic and financial powers pose a barrier to change with great obstinacy. The institutional systems still operate according to national and international priorities models inherited from the last century. Public funds granted in Europe and in different ways in other countries do not correspond to sustainable production. They very often have produced unsustainable effects with indiscriminate support to agricultural revenues. The mode of delivery of the so-called decoupled aid have product and produce financial consolidations inversely proportional to the size of the companies, social conditions and do not become carriers of new virtuous behavior. It is no longer the time of inaction and expectations. We must put everything already existing to transform the economic and cultural paradigm focused on continuous growth in a conscious vision to live within the limits of a single planet, to reverse the rapid anthropogenic transformation of the Earth and help create a truly sustainable future for all human societies. The answer to the epochal crisis cannot be increasingly aggressive conflicts and violence to attract increasingly poor resources of the world.

The emblem of a world clinging to the past is given by the enormous military expenses. In 2012 1.733 trillion dollars on weapons were spent. It would be enough to convert 10% of military spending to effectively tackle the above-mentioned devastating phenomena.

Finally, greater sustainability is very difficult but not impossible, and can only be achieved through the construction of an alternative strategy to the globalization of capital based on the role of local communities able to replace every corner of the Earth in intelligent bio-territories.

Acknowledgements

The authors wish to thank Enrico Pugliese, Professor Emeritus of Sociology of Work at the University of Rome, for his careful reading of the text and valuable suggestions. Furthermore the authors express special thanks to the director (Riccardo d'Andria) and the researchers (Fulvio Fragnito, Giovanni Morelli, Antonio Leone, Salvatore Purificato) of Cnr-ISAFoM and to the staff of Futuridea (Roberto Caruso, Mario Festa, Imma Florio, Antonio Iadicicco, Rossana Maglione, Alessandro Nardone, Francesco Nardone, Arnaldo Palombi, Giancarlo Postiglione, Simone Razzano, Roberto Romano) for the constructive willingness to discuss technical and scientific solutions on the subject of sustainability.

References

1) A. Alessandrini, Il tempo degli alberi (Ed. Abete, 1990).
2) C. Nardone, L'agricoltura italiana nel contesto internazionale; nuovi possibili strumenti di governo. Workshop 14-02-1997 dell'Accademia dei Georgofili, in: I Georgofili – Quaderni 1997 (Studio Editoriale Fiorentino, Firenze, 1997).
3) ONU Conference on Environment and Development (Rio de Janeiro, 1992).
4) European Sustainable Cities and Towns Conference (Aalborg, 1994, Lisbon, 1996, Hannover, 2000).
5) World Summit for Sustainable Development (Johannesburg, 2002).
6) C. Nardone, Crisi e sostenibilità (Il Bene Comune, 2010).
7) C. Nardone, Cibo Biotecnologico (Hevelius, 1997).
8) J. Lovelock, The Revenge of Gaia: Why the Earth Is Fighting Back – and How We Can Still Save Humanity (Santa Barbara (California): Allen Lane, 2006).
9) N. Stern, The Stern Review on the Economics of Climate Change (Cambridge University Press, Cambridge, 2006).

10) F. Cavalli Sforza and L. Cavalli Sforza, Perché la scienza. L'avventura di un ricercatore (Ed. Mondadori, 2007).

11) L. Brown, State of the World 2014: Governing for Sustainability (Worldwatch Institute | worldwatch@worldwatch.org 1400 16th St. NW, Ste. 430, Washington, 2014).

12) http://www.unccd.int

13) L. Conti, Riflessioni sulle condizioni di sostenibilità dell'agricoltura (Havelius, 1997).

14) B. Commoner, The closing circle. Nature, man and technology (Knopf, New York, 1972).

15) FAO Food and Agriculture Organization of the United Nations, The State of the land and water resources (www.fao.org/nr/solaw/solaw-home/en/, 2013).

16) VV.AA. Repertorio delle Innovazione Utili e delle Buone Prassi – Territori che fanno la cosa giusta. Pubblicazione realizzata con il contributo FEASR 2007 – 2013. ASSE IV Approccio Leader – Misura 4.2.1 "Cooperazione Interterritoriale e Trasnazionale". Progetto di Cooperazione interterritoriale "Territori che fanno la cosa giusta" – Fase 3 "Rete delle reti" (2014).

17) F. Roiatti, Il nuovo colonialismo. Caccia alle terre coltivabili (UBE, 2010).

18) E.S. Liberti, Land Grabbing. Come il mercato delle terre crea il nuovo colonialismo (Minimum fax, Roma, 2011).

19) E. Mingione and E. Pugliese, Rural subsistence, Migration, Urbanization and the New Global Food Regime in From Columbus to ConAgra. The Globalization of Agriculture and Food (University Press of Kansas, 1994).

20) D. Matassino L'importanza del recupero di tipi genetici autoctoni. Atti II Congresso Nazionale RIRAB (Rete Italiana per la Ricerca in Agricoltura Biologica) – IX Convegno ZooBioDi (Associazione Italiana di Zootecnia Biologica e Biodinamica) "Il contributo dell'Agricoltura Biologica ai nuovi indirizzi di politica agro-ambientale: il ruolo della ricerca e dell'innovazione" (Roma, 13 giugno 2014. I Quaderni ZooBioDi "La Biodiversità: una risorsa per la zootecnia biologica", 17–33, 9/2014).

21) L. Torri, B. Baroni, M.R. Baroni, Modified atmosphere (Food Packages Free Press. 2009).

22) J.T. Esquinas Alcázar. Interview by Maria Fonte and Latislao Rubbio in: La questione Agraria 53 (Franco Angeli, Milano, 1994).

23) A. Barone, L. Frusciante. Molecular marked-assisted selection for resistance to pathogens in tomato in: Marker-assisted Selection: Current Status and Future Perspectives in Crops, Livestock, Forestry and Fish (Food and Agriculture Organization of the United Nation. Rome, 2007).

24) L. Busch et al. Plants, Power and Profit:- Social, Economic and Ethical Consequence of the New Biotechnologies (Basil Blackwell inc. 3 Cambridge Center, Cambridge, Massachusetts USA, 1991).

25) C. Nardone relatore Indagine conoscitiva sulle biotecnologie, Paper presented to the italian Chamber of Deputies, Atti XIII Legislatura, (Edizioni della Camera dei Deputati, Roma, 1997).

26) VV.AA. La percezione delle aree rurali da parte dei giovani. L'agricoltura a beneficio di tutti. Documento prodotto nell'ambito della Rete Rurale Nazionale – Gruppo di Lavoro: Giovani, (MiPAAF – DISRII, 2013).

27) J. Hillmann L'anima dei luoghi. Rizzoli Milano 2004.

28) VV.AA. Desalination plants with low energy consumption and ecoserra. (UTAB Tunisian Union of Agriculture and Fisheries, Futuridea and Gssep, Tunis, 2014).

29) D.A. Vaccari. Fosforo: una crisi imminente. (Le Scienze, 2009).

30) G. Nebbia. Dove troveremo tutto il fosforo per sfamare tanta gente? (Gazzetta del Mezzogiorno, August 11, 2009).

THE "NEW" DEVELOPMENT OF RENEWABLE ENERGY SOURCES IN THE WORLD. A POTENTIAL PATH TOWARDS GLOBAL SUSTAINABILITY

GLOBAL SUSTAINABLE AND INTEGRATED DEVELOPMENT. THE CASE OF GLOBAL SUSTAINABLE AND SOCIAL ENERGY PROGRAM – GSSEP ONLUS

Carlo Sinatra

Global Sustainable Social and Energy Program – GSSEP Onlus
info@gssep.com

Abstract

The development of renewable energy sources up to now has been inspired to specific investments opportunities not corresponding to an overall and structural vision aimed to contribute to the creation of a new economic, industrial and social model. Part of the old development has been inspired to the same financial logic which provoked the current irreversible world crisis, mainly affecting now Europe but still US too. Macro drivers irreversible and of long term could give now the opportunity of a highly significant new development of renewable energy sources in a complete different logic and with the view of finally starting a path towards global sustainability, where man is really at center of the universe.

1 Introduction

The development of renewable energy sources in the world during the last years has been extremely significant, but it did not determine, actually, a correspondent social improvement and economical global growth, at least as potential and "due".

Actually, the PV cumulative installed capacity rose from 16 GW to 102GW (2012) while the wind farm capacity grew from 121 GW to 283 GW (2012)[1].

---

[1] Data Source: Bloomberg.

Such result was determined by renewable energy capacity increase in few countries and due to a particular and nonstructural institutional development frame, including a system of laws mainly based on incentives on sale price of green energy, not constructed on structural economic aspects, but just on contingent ones.

Particularly, PV installed capacity growth took place mainly in Germany and Italy, while the wind capacity grew essentially in China and US. Such development was consequently driven by contingent policy, generically aimed to contribute to the environmental world targets but was constructed on a basis which is not solid, not only from economic and industrial point of view, but also from a "social" one, and the investments done were not a concrete and harmonized part of a global and integrated development of the relevant territory, but just corresponding to specific and un-harmonized opportunities.

2   The Current Status of Renewable Energy Sources Development

From a purely economic perspective, the overall development drivers of a world portfolio of renewables sources despite such "occasional development approach" have remained robust and above notwithstanding the economic, financial and industrial frames which were mainly in Europe but not only there, being characterized by a deep and structural crisis of the industrial and economic old development and growth models.

Such crisis in the first phase of renewable energy development was (and actually still is, despite concrete signals of growth) not manageable also due to the fact that the so called new economy has not been conceived yet as a structural part of a wider development plan and all the investments done accomplished just to specific and very particular logic and reasons.

The initial, and in our view still very limited, occurring development of our economy based on renewables sources and sustainability concepts and targets, was actually mainly driven by a speculative approach, similar, *mutatis mutandis*, to the financial approach followed during last fifteen years in all the other fields, and despite objective benefits in different areas and to different players, such development did not determine any structural positive impact, or at least a significant one, on the world economy as it could have been if such development would had been driven differently and essentially by a "vision", the path to global sustainability.

Actually, the registered development in the "renewable green economy" was mainly determined by a system of laws[2] creating pure and sole tariff incentives on energy sale price, and consequently, despite important economic results for "investors", it did not create any added value in the production chain (i.e. producers of cells and turbines) up to one year ago, so that many of them, as is the case in these days also of a big Chinese cell producer, have fallen to bankruptcy or to a very negative economic situation. And all the chain was influenced negatively by such approach. Energy prices did not decrease, and on the contrary, increased so much as to block industrial development, generating increasing costs in day-to-day life in addition to further depressive impact on the economy.

No real social benefit came from such investments or they were so small to make lose their actual potentiality in as change factor of our day-to-day life.

Food and energy were mainly managed as totally alternative and the above determined an enormous quantity of opportunities lost to a real and global sustainable development approach.

But we are close to a deep and structural revolution in the overall world economic, financial and law frame, giving the opportunity to many countries of the world, mainly the emerging ones, but also, to a minor extent, to the old fashioned European ones, to construct a different development model in which renewables and more generally, energy, agriculture, and social needs satisfaction grow simultaneously, and they could be combined, or better integrated into each other successfully.

It is a real opportunity, or perhaps a need, giving our countries, no one excluded, the opportunity of a very useful and outstanding revolution, the global sustainable development one.

2.1   *The Future of Renewable Energy Development. A Real Revolution Towards Global Sustainability?*

Considering the possible future in the world of renewable energy sources and mainly of Biomass, PV, Wind, Geothermic and Hydro the questions to raise are three:

a) Is the renewable sector a field which could benefit of further development?

---

[2] With particular reference to Italy, Eastern countries (Bulgaria, Romania, Chezch Republic, Poland) but up to respectively fours and two year ago also Spain and France.

b) In case of positive answer to the above question is the renewables sector able to be a player to participate to a real and structural economic grow of our economy?

c) Is there a new way to the global sustainability goal?

No doubt that the renewable energy field would benefit in a future of a further development and actually would be probably, if we do not lose such opportunity, an outstanding development which shall deeply have influence on our style of life too for many future generations.

From a purely statistical approach we could observe that since the beginning of the year 2013 companies operating in the so called green economy and listed to Stock exchange markets grew on average adding 3 billion of Euro capitalization[3] in the first eight months of the current year, inverting a negative trend of latest three years which brought the main benefits of the incentive system on speculative investors more than on industrial producers and energy consumers. Such statistical data are probably related to a start of a new trend, but are in our view still not able to represent the width and potentiality of the coming sustainable revolution.

Particularly such new trend was determined by the progressively and wider decreasing in the world of the speculative approach (in many part of the world the law system changed significantly i.e. in Italy and in many East European countries and where a strong incentive policy was recently created, as in Japan, it was due to particular emotional and contingent factors)[4]. We strongly believe that such trend is not contingent because of being pushed by two irreversible and consequently long term macro drivers, so stable to significantly put the basis of a real change of investment approach in energy for the current and future centuries.

The first macro driver is constituted by the formal and material targets of environmental safeguard and survival, the second is determined by the energy source diversification needs.

As far as environmental issues are related, it has been currently acknowledged worldwide that sources different from renewables are actually dangerous and unsafe to the world considered in itself and to all human being as persons too. It has been scientifically and undoubtedly proven that the

---

[3] The Italian laws 5/6 July 2012 eliminated energy tariff incentives on PV and modified significantly biomass and wind related benefit; This year also Romania curt significantly the incentives, while Bulgaria and Chezch Republic did it on 2011/2012.

[4] The Japanese law system introduced by July 2012 and relevant regulations granted to METI Authorities.

combustion of fossils not only generates carbon dioxide but a large number of adverse effects.

Additionally, nuclear power appears also less and less competitive after Fukushima disaster and more and more governments and countries are inclined to manage the risks of such potential unsafely, having more and more acknowledgement also that the past envisaged economic competitiveness was actually not existing at all. The remarkable costs of latest nuclear plants make them unrealistically financeable and bankable also after the introduction of the project financing market of the so called "Project Bond". All governments in the world intend seriously to improve significantly renewables investments: sometimes, as in the Japanese case, still based on incentives policies on clean energy sales (however also the Japanese laws provide a progressively decreasing mechanism of energy sale prices) sometimes on more structural planning of their economic grow and based on the awareness that a new model of life is more and more required by populations and by a new life style to a certain extent necessary after that a pure consumer logic shall be necessarily abandoned.

On the other side the shale gas impact (not small in any case) seems really confined to US, China and Canada for many reasons, first of all, once again, the environmental one, related to its related transportation risks, and in no case it shall constitute a serious and competitive obstacle to the renewable energy sources growth; it appears more probably as a partial substitute of part of the declining "conventional sources" than an alternative to renewable energy sources.

It is no more just a question of international treaties and worldwide conferences and/or state conventions fixing long term goals and targets, now it has been starting more capillary mechanism of "laws restructuring" including those affecting foreign investments in emerging countries.

Also those countries which in the last twenty years, for the purpose of maintaining an important (but just apparent and artificial competitive factor to conventional industry and energy system) had a very low interest to environmental safeguard, have now acquired a more mature acknowledgment of the need to facilitate a grow system strongly related to environment safeguard.

The second driver, as stated above, is related to the concept of energy diversification: green economy (and technologies) may potentially benefit from unlimited resources (despite fossils ones); the more their technologies improve, the more their cost is falling down. We are still just at the beginning of such process. New and revolutionary PV cells are expected shortly, turbines and more generally efficiency combined to a technical simplification is coming and when energy could be stored.

The integration sources need shall always, as now, represent the rule of energy satisfaction system, but the mix shall significantly change and the increasing competitiveness and technical reliability should determine in the next fifty years a complete change and inversion in such a mix, above all when the sun and wind energy, as well as that produced by other natural factors, would allow for storage and not lost. Without such lost, the renewable energy produced in the world would be already now, before than the just started technological improvement would come to the expected results determining also a further significant increasing of green energy produced, more or less double the current one. Such result has to be however encouraged because of the enormous increasing need of energy itself, both related to demographic growth and to the need of accompanying emerging economies in their structural development.

All the above shall consequently procure a new industrial approach to renewables by producers themselves, strongly related to quality and technology improvement which shall determine not only positive results in their balance sheets but in all the related chain, including the "satisfaction" of end users and populations themselves also from the environmental point of view. Actually such improvement shall also facilitate a quickest result in meeting global climate change objectives.

3   The Global Sustainable Development and the Crisis of the Current Economic, Financial and Law Frame System

The big revolution coming is offered by the opportunity of substituting the old economy model, completely failed and fallen in an irreversible crisis, with a global sustainable model based on the strong relation, better integration, between energy, food and other main sources potentially determining a sustainable approach to worldwide growth. The case of Global Sustainable Social and Energy Program – GSSEP Onlus – an ONG promoting all over the world integrated big scale projects granting actual benefits to all the players wherever territories are open accepting their new global sustainable development model. GSSEP Onlus project in Tunisia, called Jasmine, an example of future growth.

But the most important consequence of the above described trend shall be the worldwide extension of green development, or better, of sustainable development, to countries different from those which currently have been developing the green economy. Actually, already now, more and more countries are supporting renewables energy (and this is also and mainly the case of emerging countries and economies in Africa, South America and Middle east) where renewables could be seriously combined to unique opportunities offered by the irreversible crisis of the current economic and industrial model of growth to be a concrete step towards the resulting global sustainability. In other words the current macro drivers could conduct ourselves to the degrowth of unsustainable growth and to the development of the global sustainable revolution.

It is a chance not to lose.

It is in fact clear to all that the competitiveness of a renewable global strategy also depends on the market and institutional framework within they operate, and that the wideness of such development it shall also be significantly influenced, both positively and negatively, by the concrete capacity of Institutions, Governments, Stakeholders, Nonprofit Associations and more generally opinion leaders. It is a question of global development of any relevant and involved territory, such a village, a town, a Province, a Region, a State: the World.

It appears consequently necessary to harmonize the International and national current frame law system to such new opportunity, definitively abandon a policy a pure energy sales tariff incentives and on the contrary improving a planning system aimed to facilitate the integration of a global development incentivizing an active role and participation of guesting territories, connecting agricultural growth to that of energy and accelerating the process (in such a case incentives both direct and indirect such as those on taxation could be more appropriate) and in that frame also make much more challenging the current environmental and social targets of European regulations as well as those of the current international treaties which could be modified with a new vision.

The renewable energy development will be more and more induced to abandon a purely speculative logic per se and shall be more and more connected to a related and integrated development of the relevant specific territory.

It appears consequently necessary to harmonize the International and national current frame law system to such new opportunity, definitively abandon a policy a pure energy sales tariff incentives and on the contrary improving a planning system aimed to facilitate the integration of a global development incentivizing an active role and participation of guesting territories, connecting agricultural growth to that of energy and accelerating the process (in such a case incentives both direct and indirect such as those on taxation could be more appropriate) and in that frame also make much more challenging the current environmental and social targets of European regulations as well as those of the current international treaties which could be modified with a new vision.

A new vision still more global and green, based on the acceleration which the new technology shall bring and energy storage shall procure.

Now also the emerging countries are ready to accelerate such process and to follow a different vision for their development: no more growth based on old economy models, which dramatically failed, but on the contrary also revisiting their growth targets in quantity based on integrated growth with a very long term approach.

It is time stopping development models also indirectly related to the alternative growth of energy and food.

Both represent the most outstanding need of our populations and should be developed not only simultaneously but also in an integrated way. We must abandon the alternatives which old biomass technology has procured and also at international level a new economic and institutional global model has to be facilitated avoiding the proliferation of international treaties fixing just very long term targets isolated each other, establishing, on the contrary, more concrete and planned targets goals, integrated between them.

4   Global Sustainable Social and Energy Program – GSSEP ONLUS

Global Sustainable and Social Energy Program – GSSEP Onlus – is a nonprofit (ONG) organization formed in Italy on March 2013 with the purpose of concretely promoting dialoguing with Governments and Institutions big scale integrated projects based on the combined development of renewable energy and agriculture and the simultaneous and related implementation of social investments (schools, hospitals, multicultural center, human rights protection). More particularly, GSSEP Onlus developed a model which combines growth and availability of food, energy and quality of life. Such model is based on the assumption which all the "involved parties" i.e. industrial players, more generally industrial investors, farmers, guesting populations, guesting authorities and Governments and finally GSSEP Onlus itself have to take significant benefit from the "integrated project". Practically part of industrial profits of all the contemplated renewable investments upper a certain level are utilized to support and sustain the agriculture business and giving it reasonable profitability as well to realize through GSSEP Onlus social investments for the benefit of the territory net of the levied costs with a royalties systems totally dedicated by GSSEP Onlus to the implementation of the social plan:

GSSEP's main concrete principles are:

(1) Keep the main part, and possibly the whole, of energy output in the territories where the initiatives are implemented, so to promote local growth and development. GSSEP's renewable energy sources are mainly biomass of second generation (no use of food but just of the alimentation scraps and crops/residues) combined to a compost system so to not deteriorate the quality of soil, related biogas and thermic energy, photovoltaic, hydro power, geothermal and wind energy plant, all or part of them (as the territory allows but always with biomass second generation important plants)combined into dedicated "agro and bio energy islands" adapted to the physical characteristics of the territories. Such large agro bio energy islands are of very significant dimensions and contribute to a massive production of energy in the frame of a fully integrated sustainable "scheme". Such island are also characterized by new clean, innovative and globally sustainable technologies developed by high qualified associates of GSEEP Onlus in agreement and coordination with GSSEP Onlus itself.

(2) Promote the creation of new food and alimentation as a value for itself and also to generate crops and residues to be dedicated to the second generation biomass plant.

(3) Generate very significant new employment in agriculture (each GSSEP Onlus agro bio energy island could bring at least six thousand new farmer) and namely in young people (possibly less than thirty years old) and women consortia; develop more generally permanent employment and professional improvement and training also on the "new technologies" related to GSSEP Onlus projects.

(4) Implement civil and social infrastructure with part of the profit generated by the industrial renewable investments. Actually all of GSSEP's initiatives are necessarily associated to very important infrastructural investments such as improvements, upgrading and reinforcement of roads, railways, airports and ports, as needed, but are also focused on many social investments in schools, hospitals, multicultural centers, renewable energy "living villages and museums", childhood and woman organizations, training and education programs on food, energy and sustainable development targets.

(5) Generate a win/win new economic system and cycle in the territory aimed to supersede the crisis of the old current model and concretely generating a global sustainable energy and social program.

(6) Promote the environmental protection of the territories and of clean renewable energy and recycling of waste too; reducing pollution.

(7) Develop joint scientific research in the fields of sustainable agriculture production and use of energy.

(8) Set up permanent bilateral platform of cooperation with interested governments and institutions, as well as with main stakeholders and other ONG organizations also enhancing coordination between environmental organizations in different countries.

Currently GSSEP's organization is based in Italy but it is intended to create branches all over the world and being organized as a worldwide initiative with the objective of promoting GSSEP Onlus principles in all countries focusing on the implementation of sustainable growth for their populations.

GSSEP Onlus is a new philosophy to address development and cooperation, based on values of professional competence, transparency, ethical and deontological approach and the global sustainable vision aimed to change completely the current economic cycle and taking opportunity of the current world crisis to "propose" a new and fully sustainable development.

GSSEP Onlus is led by a group of experience professionals, actively engaged in many disciplines such as development, legal, economy, finance, energy, agriculture, social responsibility, environmental science, engineering, chemistry and public affairs.

A broad program has been established to start concretely the implementation of this overall world change program. Currently the starting points have been identified in the following countries: Tunisia, China, Laos and Cambodia, Kazakistan.

5   The JASMINE Project (GSSEP ONLUS Project for Tunisia)

GSSEP Onlus has elaborated with the support of high qualified researchers and its own associates a global project for Tunisia, called Jasmine, based on the potential development of six bioenergetics islands ubicated in the following Regions: Kebili, Tataouine, Medenine, Keirouane, Kasserine and Sidi Bou Zid.

The configuration of each island for Tunisia has been defined, provided that the single bio energy island could vary mainly due the very different characteristics of soil and water and that the agricultural plan, so essential to determine the dimension of the biofuel one, shall also vary a lot.

Indicatively each of the bio energy island shall be composed by:

(1) a biomass/biofuel plant alimented just by food crops and residues and not by food itself (biomass of second generation) having a maximum capacity of 13 MW, as well as by a related compost plan;

(2) a PV plant having a capacity not lower than 40 MW;

(3) related biogas and thermic energy production plants;

(4) a massive new agricultural activity, changing in each "bioenergy island" as a matter of cultivated food and plantations, aimed to generate new massive food to remain in the Guesting territory and to produce the crops necessary for the biomass production; and

(5) a social plan taking place through the creation of social investments, such schools, hospitals, multicultural centers, specific activities for the protection of human rights, childhood and women emancipation.

Both in biomass and in agriculture production the project is characterized by innovative technology and by a full sustainable approach granting a very material improvement of the environment too, additional to reasonable profits for industrial investors, creation of relevant quantities of clean energy and food, significant new employment, direct and indirect growth and social/civil needs satisfaction.

All the above based on a very detailed economic model and related business plan, governing in an integrated way the results of each "business", i.e.: the renewable energy (industrial), the agricultural and the social ones.

More particularly the business plan provides that the biomass/biofuel plant produces bioethanol to be sold in Europe so getting advantage of the current UE laws and regulations granting to such product an incentivized tariff which determines from an economic point of view for their producers a robust profit. Part of this (potential) profit, obviously when generated, shall be "diverted" partially as a granting the agricultural cooperative selling the crops at a price above the market one so to contribute to agricultural costs mainly connected to water availability systems and partially through an initial lump sum and year to year royalties to be dedicated to implement the social plan. Also GSSEP Onlus receives an amount by the industrial players as voluntary contribution connected to its model conception, management, advisor and active following up of the project, amount which, net of living costs levied and to be levied in the Project, is reinvested in the social plan or, as will be in Tunisia, in an wide and articulated multicultural centers which GSSEP Onlus intends to implement in the desert of Tataouine.

The current not yet formalized agreement between GSSEP Onlus and the Tunisian Government provides that the industrial activities of each bioenergy island be object of an international bid process where the entire envisaged industrial plants be considered jointly based on a predetermined availability of crops and other bankable conditions, as hereinafter indicated.

On the contrary the agricultural activities shall be organized by the reorganization of part of existing one and through the creation for each bio energetic island of indicative 4000/5000 ha of cultivation through the creation of agricultural consortia/cooperatives, participated also by the Tunisian Government (Ministry of Agriculture and UTAP[5]) which shall enter into crops sale and purchase agreement "guaranteed" for the period of industrial business by the Government itself so to give bankability of the biomass supply for the period necessary (25 years) and at terms and conditions which would allow the industrial player getting financed.

Under GSSEP Onlus model and mission such cooperative/consortia should be mainly composed by young people (possibly under 30 years) and by women so simultaneously giving a concrete contribution to the current young deployment in the world to women emancipation in the territory.

Each island should determine at least 7000 new farmers, while the industrial activity should imply a further direct employment of about 150 employees additionally increased from indirect employment (transportation activities) not yet precisely valuable and depending on different issues.

The social plan shall be implemented progressively based on a mathematic royalty system which should determine at least for each bio energy island the creation of one hospital, three childhood centers, 2 schools, one multicultural center (theatre, music, dance, arts).

In consideration of the innovative technology approach granted to GSSEP Onlus associates the project also provides locally free training courses. Particularly, due to an agreement entered on 27/28 November 2013 between GSSEP Onlus and Utap specific innovation systems shall be introduced in the agriculture activities so make them absolutely unique, currently, as a matter of sustainable development giving the agricultural project a worldwide characterization.

The advantages of Jasmine for Tunisia are many and very important.

Investments on The Territory in concrete and productive activities of more than 1,5 billion Euro, a significant quantity of clean energy and food available for the population and its civil and social growth, massive development of an

---

[5] UTAP Union Tunisienne de l'Agriculture et de la Pêche.

environmental and safe plan which constitute a requalification of six Region including those two in the desert (Kebili and Tataouine), a global employment of at least 50.000 people, new social investments free of charge and technology improvements both in industry and in agriculture.

GSSEP Onlus shall act formally as Tunisian Government advisor and shall prepare in agreement with the same the industrial project frame in relation to which to offer in the international bid process as well as the agricultural project in agreement with the Government and Utap assisting them as international advisor, jointly with its own associates and researchers. GSSEP Onlus shall also manage in agreement with single regional Governments the social plan acting as independent guarantor of its implementation.

All the costs of GSSEP Onlus and its activity shall be considered and paid by the winner of the international bid and shall be acknowledged for all the period of construction of the industrial plants and implementation of the social plan.

The frame system necessary to implement such a big integrated project needs naturally being tailored and that means that specific laws and regulations would be required be put in place to make the Bid successfully.

More precisely the Winning Bidder should be provided:

(1) with legal instruments (concessions for 25/30 years at symbolic fee) to grant the land necessary to implement the industrial activities;
(2) a tax/fiscal exemption system for a period of ten years on all the Project revenues;
(3) an integrated and special authorization process consistent with the Tunisian law system and major environmental and sustainable principles but to be completed in a period not exceeding six/nine months from the date of starting the authorization process/the formal contractual award;
(4) a specific tariff scheme to be provided for all the produced power (PV and biogas) which shall be bought by the Government/Steg. Such tariff shall be proposed by the Bidders and shall constitute an element of evaluation of the Project provided that in order not to deprive the social investment plan it shall be offered having in the offer itself a minimum royalty plan which the Tunisian Government and GSSEP Onlus shall agree inserting in the Bid scheme for each single Region; and
(5) full support and planning, whenever necessary, to grid adaptation by STEG/the Tunisian Government.

Global Sustainable projects need global sustainable frame and a law system matching the big opportunity offered by big scale integrated projects.

New specific and "global" rules are needed substituting the two much fragmented current laws and regulations.

6   Conclusion

Integration between renewables, agriculture and social represents the future and must be accompanied by a new social acknowledgment not based on consume needs increasing but on the contrary on consume needs optimization and requalification.

Renewable biomass of second generation sources could be generated by waste and crops of food and are not consequently alternative to them but on the contrary could generate a further opportunity of earning for farmers and significant (millions in the world) employment in agriculture.

Time has come to construct a new global model to ensure all the countries and also the developing ones, real access to sustainable, reliable and affordable energy, not affecting climate change and taking into account economic, technologic and geopolitical issues, simultaneously increasing very significantly on the earth food and alimentation safe production. Time is come to bring at the center of this new model not the capital and or the finance but ourselves, the humanity and our global sustainable, economic and social, development.

An integrated global sustainable energy, agriculture and social plan granting a winning result to all the players, (industrial ones, agricultural ones, local and national governments, populations, ONG promoters too) is more than possible and can be concretely implemented as it witness the special business plan and integrated economic model prepared by GSSEP Onlus[6].

The above concepts implies however also a deep rethinking of the system of law applicable and needs an integrated specific regulation case by case based on two essential macro drivers: 1) granting not speculative but necessarily industrial incentives system (bankable ones) in order to allow the industrial players to participate successfully to the project, keeping reasonable industrial projects and actually enlarging part of them upper the above level so to sustain the agricultural business and finance the social associated plan 2) a large scale sustainable and global development, fully consistent with environmental and social needs finally acting as a real change factor on social and civil development, above all in emerging countries.

Finally, if so properly constructed, the Project (the industrial one) should also benefit of primary national and international bank interest based on the

---

[6]   The GSSEP model was presented at OECD National Forum on April 2013.

above said bankability under a scheme of project financing, condition which could determine a reasonable quick expansion of GSSEP Onlus project concept, hopefully in many other parts of the world, not losing the big current opportunity of a real change for the benefit of all of us.

A PROPOSAL FOR ADVANCED SERVICES AND DATA PROCESSING AIMING AT THE TERRITORIAL INTELLIGENCE DEVELOPMENT

Salvatore Rampone, Gianni D'Angelo
University of Sannio
Dept. of Science and Technology
Benevento, Italy
{rampone, dangelo}@unisannio.it

Abstract

Intelligent communities and territories belong to an emerging movement targeting the creation of better environments. Technological information is recognized as an important factor shaping territorial systems of innovation. This paper focuses on territorial intelligence: distributed information systems localized over a region allowing continuous update and learning on technologies, competitors, markets, and the environment. Namely we describe the general problems of information extraction from raw data and automatic construction of knowledge bases from databases. Then we introduce a learning algorithm which integrates machine learning methodologies, databases technologies and high performance computing in order to discover the laws that govern an environmental process starting from examples of the process behavior itself.

1 Introduction

Territorial intelligence is the science having for object the sustainable development of territories and having for subject territorial community [1]. Intelligent communities and territories belong to an emerging movement targeting the creation of better environments [2]. Technological information is recognized as an important factor shaping territorial systems of innovation [3]. This paper focuses on territorial intelligence: distributed information systems localized over a region allowing continuous update and learning on technologies, competitors, markets, and the environment.

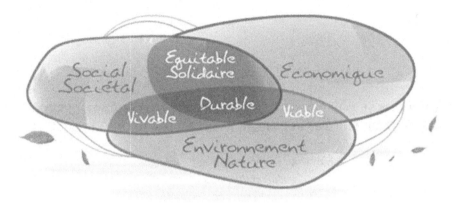

Figure 1. A classical schema of territorial data integration. Such data come often from various sources yet unstructured such as social media, sensors, scientific applications, Internet, medical records and business transactions.

Some years ago the International Organization for Standardization proposed the ISO 14000 family, addressing various aspects of environmental management. It provides practical tools for companies and organizations looking to identify and control their environmental impact and constantly improve their environmental performance. Specifically ISO 14001:2004 and ISO 14004:2004 focus on environmental management systems [4].

However, implementing an environmental management system can be challenging. Specifically the great amounts of data produced from many sources has led to the new challenges of the Big Data [5]. Big Data are the collection of large amounts of unstructured data with fast input/output, such large that it is difficult to capture, manage, process and model the data within a tolerable elapsed time. Such datasets come often from various sources yet unstructured such as social media, sensors, scientific applications, Internet, medical records and business transactions.

Because of its massive data sets, today environmental sustainability meets Big Data. The problem is not only data but interpretation [6].

Data have hidden information in them and to extract this new information, interrelationship among the data has to be achieved. So, Big data is more than a size and storage issue; it is an opportunity to find insights new and emerging types of data and content, and to answer questions that were previously considered beyond our reach. The main challenges in handling Big Data lie not only in the four V's [7], namely, huge Volume in amount, high Variety in type, Velocity in terms of real-time requirements, and Variability, i.e. constant changes in data structure and user interpretation, but also in the approach to understanding data through methods able to classify the data and find a suitable pattern among them. Data analysis techniques that have been traditionally used include regression analysis, cluster analysis, numerical taxonomy, multi-dimensional analysis, multivariate statistical methods, stochastic models, time series analysis, nonlinear estimation techniques, and others [8]. These techniques are primarily oriented toward the extraction of quantitative and statistical data characteristics, and as such have inherent limitations. For example, a statistical analysis can determine correlations between variables in data, but cannot develop a justification of these relationships in the form of higher-level logic-style descriptions and laws. In the numerical taxonomy technique, it is possible to classify organisms by a comparison of large numbers of observable characteristics, but it isn't possible to get the qualitative description of the classes created and cannot give reasons why the entities belong to the same category. To address such issues, a data analysis system should be equipped with a substantial background of knowledge and, then, being able to perform symbolic reasoning on the data. Yet, these interpretations and insights are just the goal sought by those who build databases. So, the standard data analysis methods need to be extended to get the information and extract new knowledge. To overcome the above limitations, researchers have turned to ideas and methods developed in machine learning [9], which goal is to develop computational models for acquiring knowledge starting from facts and background knowledge. These and related efforts have led to the emergence of a new research area, frequently called "Data Mining (DM)" and "Knowledge Discovery in Databases (KDD)" [10]. KDD is the organized process of identifying valid, novel, useful, and understandable patterns from large and complex data sets. Data Mining is the core of the KDD process, involving the inferring of algorithms that explore the data, develop the model and discover previously unknown patterns. Data mining consists of many up-to-date techniques such as classification (decision trees, naive Bayes classifier, k-nearest neighbor, neural networks), clustering (k-means, hierarchical clustering, density-based clustering), association (one-dimensional, multidimen-sional, multilevel association, constraint-based association). Nevertheless, the user with his knowledge and intuition about the application domain should participate in the search for new structures in data, e.g. to introduce a priori knowledge and to guide search strategies. The final step in the inference chain is the validation of the data where new techniques are called for to cope with the large complexity of the models.

Various technologies are being discussed to support the handling of big data such as massively parallel processing databases, scalable storage systems, cloud computing platforms, and others [11]. Architectures for big data usually range across multiple machines and clusters, and they commonly consist of multiple special purpose sub-systems. Then, due to the difficulty of analyzing such large datasets, big data presents also a systems engineering and architectural challenges.

In this paper, we describe the general problems of information extraction from raw data and automatic construction of knowledge bases from databases. Then we introduce a learning algorithm named U-BRAIN (Uncertainty-managing Batch Relevance-based Artificial INtelligence) which integrates machine learning methodologies, databases technologies and high performance computing in order to discover the laws that govern an environmental process starting from examples of the process behavior itself [12].

The remaining of this paper is organized as follows. Section 2 is an overview of the Knowledge Discovery in Databases process and the Taxonomy of Data Mining methods. In section 3, we describe the learning algorithm. The High Performance Computing oriented Parallel implementation of the algorithm is described in section 4. Last section is devoted to the conclusions.

2   Knowledge Discovery Overview

2.1   *KDD*

As depicted in Figure 2, the Knowledge Discovery in Databases (KDD) consists of many steps iterative and interactive. The KDD process is iterative at each step, meaning that may be required to come back to previous steps. The process starts with determining the KDD goals which prepares the scene for understanding what should be done, and ends with the implementation of the discovered knowledge.

Having defined the goals, in the Selection step the data that will be used for the knowledge discovery should be determined. This includes finding out what data is available, obtaining additional necessary data, and then integrating all the data for the knowledge discovery into one data set, including the attributes that will be considered for the process. This process is very important because the Data Mining learns and discovers from the available data. This is the evidence base for constructing the models. If some important attributes are missing, then

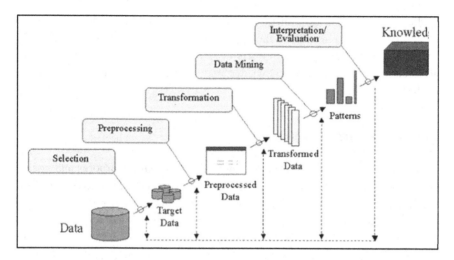

Figure 2. The Process of Knowledge Discovery in Databases.

the entire study may fail. In the Preprocessing stage, data reliability is enhanced through data clearing, handling missing values or noise. This stage may involve complex statistical methods or others Data Mining algorithm. Data Transformation step is used to generate better data for the data mining. Methods here include dimension reduction (such as feature selection and extraction and record sampling), and attribute transformation (such as discretization of numerical attributes and functional transformation). Data Mining stage represents the tactic to be used for searching knowledge, for example, classification, regression, clustering and others. Choosing mostly depends on the KDD goals and on the previous steps. However, there are two major goals in Data Mining: prediction and description [13]. Prediction is often referred to as supervised Data Mining, that is the variables under investigation are split into two groups: explanatory variables (input attributes) and one or more dependent variables (output attributes). The target of the analysis is to specify a relationship between the explanatory variables and the dependent variable. Descriptive Data Mining includes the unsupervised and visualization aspects of Data Mining, that is no target variable is identified as such. Descriptive modeling is a mathematical process that describes real-world events and the relationships between factors responsible for them. Most data mining techniques are based on inductive learning, where a model is constructed explicitly or implicitly by generalizing from a sufficient number of training examples. The underlying assumption of the inductive approach is that the trained model is applicable to future cases. In the Evaluation stage the mined patterns (rules, reliability etc.), with respect to the goals defined in the first step, are evaluated and interpreted. Moreover, the effects of the choices of the previous steps are also considered. Finally, the knowledge may be incorporate into others system for further action.

2.2   *Data Mining Taxonomy*

There are many methods of Data Mining used for different purposes and goals. As it is shown in Figure 3, the Data Mining is distinguished between two main types: verification-oriented (the system verifies the user's hypothesis) and discovery-oriented (the system finds new rules and patterns autonomously). The discovery method branch consists of prediction methods versus description methods. Descriptive methods are oriented to data interpretation, which focuses on understanding (by visualization for example) the way the underlying data relates to its parts. Prediction-oriented methods aim to build a behavioral model, which obtains new and unseen samples and is able to predict values of one or more variables related to the sample. Most of the discovery-oriented Data

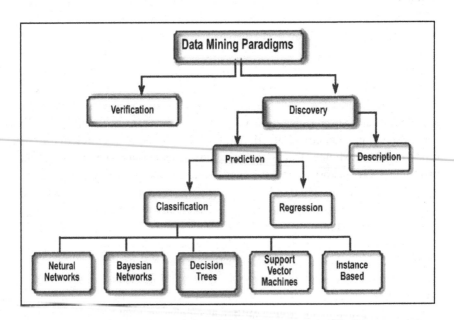

Figure 3. Data Mining Taxonomy.

Mining techniques (quantitative in particular) are based on inductive learning, where a model is constructed, explicitly or implicitly, by generalizing from a sufficient number of training examples.

Verification methods, on the other hand, deal with the evaluation of a hypothesis proposed by an external source like an expert. These methods are less associated with Data Mining than their discovery-oriented counterparts, because most Data Mining problems are concerned with discovering an hypothesis rather than testing a known one. Another common terminology, used by the machine-learning community, refers to the prediction methods as supervised learning, as opposed to unsupervised learning. Unsupervised learning refers mostly to techniques that group instances without a specified dependent attribute. Supervised methods are methods that attempt to discover the relationship between input attributes (sometimes called independent variables) and a target attribute (sometimes referred to as a dependent variable). The relationship discovered is represented in a structure referred to as a model. It is also useful to distinguish between two main supervised models: classification models and regression models. Classifiers map the input space into predefined classes on the basis of a training set of data containing observations whose category membership is known. Typical examples include, support vector machines, decision trees, probabilistic summaries, or algebraic function. On the other hand the regression models determine the linear relationship between two or more variables in order to predict a dependent variable or response from a number of independent or input variables.

3 U-BRAIN Algorithm

The U-BRAIN (Uncertainty-managing Batch Relevance-based Artificial INtelligence) algorithm [12] is a learning algorithm that finds a Boolean formula (f) in disjunctive normal form (DNF) [13], of approximately minimum complexity, that is consistent with a set of data (instances). The conjunctive terms of the formula are computed in an iterative way by identifying, from the given data, a family of sets of conditions that must be satisfied by all the positive instances and violated by all the negative ones; such conditions allow the computation of a set of coefficients (relevances) for each attribute (literal), that form a probability distribution, allowing the selection of the term literals. This algorithm, in its first implementation (named BRAIN), was originally conceived for recognizing splice junctions in human DNA [14]. Splice junctions are points on a DNA sequence at which "superfluous" DNA is removed during the process of protein synthesis in higher organisms. The general method used in the algorithm is related to the STAR technique of Michalski [15], to the candidate-elimination method introduced by Mitchell [16], and to the work of Haussler [17]. The BRAIN algorithm was then extended by using fuzzy sets [18], in order to infer a DNF formula that is consistent with a given set of data which may have missing bits.

The algorithm great versatility, as evidenced by numerous applications [14,19–20], makes U-BRAIN suitable for the analysis and integration of our territorial heterogeneous data.

UBRAIN models a process starting from a limited number of features of interest from examples, data structures or sensors. In the world of computational biology, where the amount of data grows rapidly due to new sequencing techniques (NGS, microarrays, etc.), U-BRAIN responds with the ability to learn quickly with readily available computing resources. Data base as HS3D [21], COSMIC [22] represent typical data processed by U-BRAIN. However, the overall algorithm time complexity is $\approx O(n5)$ and the space complexity is in the order of $\sim O(n3)$ for large n (with n the number of variables) [23,24]   and this is a strong limit in Big Data treatment.

4   HPC Oriented Parallel Implementation of U-BRAIN

In order to overcome the limitations related to high computational complexity, recently an high performance parallel based implementation of U-BRAIN has been realized [24]. Mathematical and programming solutions able to effectively implement the algorithm U-BRAIN on parallel computers have been found; a Dynamic Programming model [25] has been adopted. Finally, in order to reduce the communication costs between different memories and, then, to achieve efficient I/O performance, a mass storage structure has been designed to access its data with a high degree of temporal and spatial locality [26]. Then a parallel implementation of the algorithm has been developed by a Single Program Multiple Data (SPMD) technique together to a Message-Passing Programming paradigm. The speed-up of the parallel implementation varying the number of processors on HS3D dataset and COSMIC dataset are depicted in Figure 4 and 5, respectively.

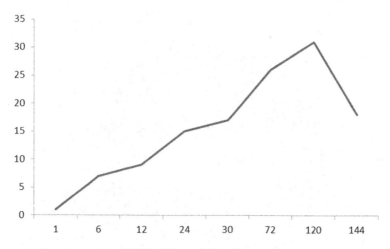

Figure 4. Speed-up of the U-BRAIN parallel implementation on HS3D varying the processor number.

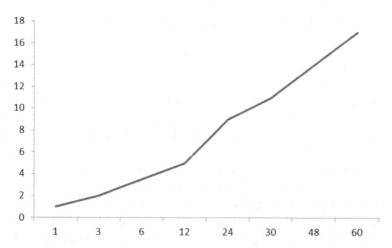

Figure 5. Speed-up of the U-BRAIN parallel implementation on COSMIC varying the processor number.

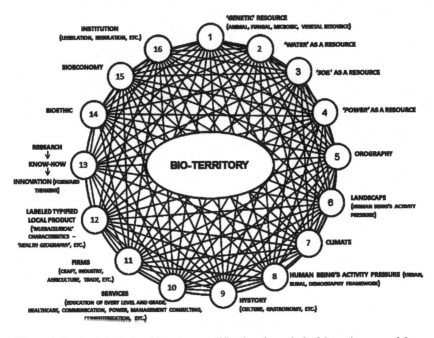

Figure 6. Territorial relationship set exemplification through the Matassino «mandala».

5   Conclusions

In this paper, the general territorial intelligence problem has been afforded from the point of view of information extraction from raw data and automatic construction of knowledge bases from databases. Data have hidden information in them, so big data provides an opportunity for big analysis leading to big opportunities to gain a competitive edge, to advance the quality of life, or to solve specific scientific issues. Successful Knowledge Discovery in Databases (KDD) and Data Mining applications play an important role in finding new meanings from the data. We introduced a learning algorithm (U-BRAIN) which integrates machine learning methodologies, databases technologies and high performance computing in order to discover the laws that govern an environmental process starting from examples of the process behavior itself. It is evidenced that sharing information between scientific disciplines, sectors and territorial scales is necessary to co-construct knowledge and collaborative projects combining economic, social, environmental and cultural objectives of sustainable development. This is the basis of intelligent distributed information systems allowing continuous update and learning on technologies, competitors, markets, and the environment. An important future application of these systems will be to master at least in part the complex territorial relationship set, well exemplified by Matassino in his "mandala" [28], reported in Figure 6.

References

1) J.J. Girardot, "Territorial Intelligence and Innovation for the Socio Ecological Transition", 9th International conference of territorial intelligence, ENTI, Strasbourg, 2010.
2) D. Matassino, M. Occidente, "Alcune riflessioni etiche sulla gestione di un bioterritorio", Italia etica Numero 1, Anno VI, 1 Aprile 2012: 8–28.
3) N. Komninos, "Regional intelligence: distributed localised information systems for innovation and development", Int. J. Technology Management, Vol. 28, Nos. 3/4/5/6, 2004.
4) http://www.iso.org/iso/home/standards/management-standards/iso14000.htm
5) S. Singh, N. Singh, "Big Data Analytics", International Conference on Communication, Information & Computing Technology (ICCICT), Oct. 19–20, Mumbai, India, 2012.
6) O. Zik, "Examining the Gap between Big Data and Environmental Sustainability Measurement", Environmental Leader. [Online]. Available: http://www.environmentalleader.com/2012/07/18/examining-the-gap-between-big-data-and-environmental-sustainability-measurement/ July 18, 2012.
7) IBM big data platform [Online]. Available: http://www-01.ibm.com/software/data/bigdata/. Last access: April 2014.
8) S. Sumathi, S.N. Sivanandam, "Introduction to Data Mining and Its Applications", Springer edition, 2006.
9) S. Michalski, J. Carbonell and T. Mitchell, "Machine Learning: An Artificial Intelligence Approach", TIOGA Publishing Co., Palo Alto, California, 1983.
10) U. Fayyad, G. Piatetsky-Shapiro and P. Smyth, "From Data Mining to Knowledge Discovery in Databases", Advances in knowledge discovery and data mining, American Association for Artificial Intelligence, Menlo Park, CA, 1996.
11) V.R. Borkar, M.J. Carey, C. Li, "Big data platforms: What's next?", The ACM Magazine for Students, Vol. 19, no.1, 2012.
12) S. Rampone, C. Russo, "A fuzzified BRAIN algorithm for learning DNF from incomplete data", Electronic Journal of Applied Statistical Analysis (EJASA), Vol. 5, n.2, pp. 256–270, 2012.
13) E. Mendelson, Introduction to Mathematical Logic. Chapman & Hall, London, 1997, p. 30.

14) S. Rampone, "An Error Tolerant Software Equipment for Human DNA Characterization", IEEE Transactions on Nuclear Science, Vol. 51, n.5, pp. 2018–2026, 2004.

15) R.S. Michalski, "A theory and methodology of inductive learning", Artificial Intelligence, Vol. 20, pp. 111–116, 1983.

16) T.M. Mitchell, "Generalization as search", Artificial Intelligence, Vol. 18, pp. 18: 203–226, 1982.

17) D. Haussler, "Quantifying inductive bias: A learning algorithms and Valiant's learning framework", Artificial Intelligence, Vol. 36, pp. 177–222, 1988.

18) L.A. Zadeh, "Fuzzy sets", Information and Control, Vol. 8, n.3, pp. 338–353, 1965.

19) S. Rampone (1998). Recognition of Splice-Junctions on DNA Sequences by BRAIN learning algorithm. Bioinformatics Journal, 14(8), 676–684.

20) G. D'Angelo, S. Rampone, "Diagnosis of aerospace structure defects by a HPC implemented soft computing algorithm", Proceedings of the IEEE International Workshop on Metrology for Aerospace, pp. 408-412 Benevento, Italy, May 29–30, 2014.

21) P. Pollastro, S. Rampone "HS3D, a Dataset of Homo Sapiens Splice Regions, and its Extraction Procedure from a Major Public Database", International Journal of Modern Physics C, Vol. 13, n.8, pp. 1105–1117, 2003.

22) S.A. Forbes, "COSMIC. mining complete cancer genomes in the Catalogue of Somatic Mutations in Cancer", Nucleic Acids Research, Vol. 39 (suppl. 1): D945–D950, 2011.

23) D. Knuth, "Big Omicron and big Omega and big Theta", SIGACT News, pp. 18–24, Apr.-June 1976.

24) G. D'Angelo, S. Rampone, "Towards a HPC-oriented parallel implementation of a learning algorithm for bioinformatics applications", BMC-Bioinformatics, Vol. 15 Suppl. 5, 2014.

25) T.H. Cormen, C.E. Leiserson, R.L. Rivest, C. Stein, Introduction to Algorithms. Boston: The MIT Press, Third edition: 2009.

26) J.S. Vitter, "External Memory Algorithms and Data Structures: Dealing with Massive Data", ACM Computing Surveys, Vol. 33. n.2, pp. 209–271, 2001.

27) W.H. Liggett, D. Sidransky, "Role of the p16 tumor suppressor gene in cancer", J. Clin Oncol., Vol. 16, n.3, pp. 1197–206, 1998.

28) D. Matassino "Global sustainability for a world of 'smart' bio-territories" in Proceedings First International Workshop "Global Sustainability Inside and Outside the Territory", Benevento, 14 February 2014, World Scientific.

VISIBLE-NEAR INFRARED REFLECTANCE SPECTROSCOPY FOR FIELD SCALE DIGITAL SOIL MAPPING. A CASE STUDY

Antonio P. Leone*, Fulvio Fragnito, Giovanni Morelli, Maurizio Tosca
Consiglio Nazionale delle Ricerche,
Istituto per i Sistemi Agricoli e Forestali del Mediterraneo
Via Patacca, 85 Ercolano (NA)

Natalia Leone
Università degli Studi del Molise, Dip. Di Bioscienze e Territorio
C.da Fonte Lappone – 86090 Pesche (IS)

Massimo Bilancia
Dipartimento di Informatica, Università di Bari Aldo Moro
Via E. Orabona 4, 70125 Bari

Maria Luisa Varricchio
Research Staff Member Futuridea
C.da Piano Cappelle, 82100 Benevento

Abstract

The aim of this work is to present a method for "intelligent", field-scale digital soil mapping based on visible-near infrared (vis-NIR) reflectance spectroscopy, in combination with statistical analysis (Principal Component Analysis, PCA and geostatistics). The study was carried out in a site of southern Italy. With reference to a 50 × 50 cell size grid, 240 soil samples were collected to a depth of 20–30 cm. The soil was analyzed by vis-NIR reflectance spectroscopy and the data were decomposed by PCA. The first three components (PC1, PC2, PC3) explained 98% of the total variance of the initial data set and therefore they were selected for the assessment of soil spatial variability by variography and kriging (geostatistics). The resulting PC1, PC2 and PC3 kriging maps were interpreted in the light of the information contents on reflectance spectra and compared with the results of a previous, conventional soil survey. The presented strategy seems to be efficient and reliable to use, when mapping soil spatial variability.

---

* Corresponding author

1 Introduction

Soil consists of a heterogeneous mixture of mineral particles of different sizes derived from weathered rock and sediments, organic particles consisting of organic compounds in different stages of decomposition, living organisms, water, air and different chemicals such as mineral nutrients associated with the particles or the soil solution. Soil-forming factors such as parent material, climate, topography, vegetation and human impact have over time resulted in a variety of different soils over the world.

The soil-forming processes also work on a regional scale, sometimes dividing farms or fields into parts with very different soil types. Together with processes acting on local field scale, this results in sometimes very large variations in soil properties within a seemingly homogeneous field [1]. This has consequences for plant production, since the growing conditions can differ over the fields in terms of plant nutrients and water availability [2,3].

Both economic and environmental benefits can be achieved by adapting soil management to within-field variability (site-specific or precision farming) [4].

One way to investigate soil variation would be to produce soil maps based on a vast number of traditional soil analyses, such as chemical, biological and physical, but such analyses are both time-consuming and expensive [5].

An alternative method to assess field soil variation would be the use of visible-near infrared (vis-NIR) reflectance spectroscopy. Reflectance spectroscopy refers to the measure of spectral reflectance [6] i.e., the ratio of the electromagnetic radiation reflected by a soil surface to that which impinges on it [7]. In vis-NIR spectroscopy, soil samples are scanned over the entire vis-NIR region (350–2500 nm) by use of a spectroradiometer. Absorptions of electromagnetic radiation in this range provide diagnostic measures of the chemical, physical and mineralogical composition of the soil (e.g. Clark *et al.*, 1990 [8]; Viscarra Rossel *et al.*, 2010 [9]). A vis-NIR spectrum provides an integrative measure of the soil. It contains information on its color, its iron oxide, clay and carbonate mineralogy, its organic matter content and composition, the amount of water present and its particle size [9].

Because vis-NIR spectra provide an integrative measure of the soil, they have been used, in combination with statistical analysis, as a fast and inexpensive (then "intelligent") tool for field-scale mapping soil spatial variability [5,10]. However, although powerful, the vis-NIR reflectance spectroscopy approach to soil mapping is still poorly understood by farmer and agricultural technicians. A need, then, exists to increase dissemination of knowledge, about this approach.

This paper aims to give an answer to this need, by presenting a concrete case study. The focus will be primarily on the illustration of a methodological approach rather than the specific results. That approach, will combine vis-NIR reflectance spectroscopy with statistical analysis, namely principal component analysis (PCA) and geostatistics. Firstly, the vis-NIR data set was decomposed by PCA, and secondly, the scores for the first three components of PCA (PC1, PC2 and PC3) were used to create spectral soil maps.

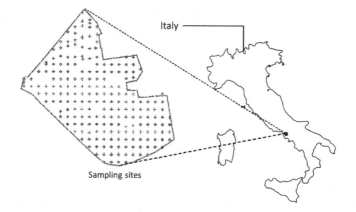

Figure 1. Location and sampling points of the investigated field in Italy.

2   Material and Methods

2.1   *Site Description and Sampling*

The field chosen for the study is located in the north-west part of the Campania Region, Sothern Italy (Fig. 1), within an abandoned meander of the lower course of Volturno river, near the municipality of Capua (Province of Caserta). Land use is dominated by fruit trees and cereals. The main soil types [11] are *Haplic* and *Fluvic Cambisols, Haplic Luvisols* [12].

Within the field site, the centre of a 50 × 50 m cell of a grid was marked with help of a GPS, making a total of 240 points (Fig. 1). Soil samples were taken with the help of an auger to a depth of 20–30 cm. The samples were air-dried and ground to a size fraction of 2 mm. Before the vis-NIR analysis, the dried and sieved samples were ground with mortal and pestle.

2.2   *Vis-NIR Spectroscopy and Spectroscopic Analyses*

The diffuse vis-NIR reflectance of soil samples was measured in the laboratory, under artificial light, using a FieldSpec Pro spectroradiometer (Analytical Spectral Devices, Boulder, Colorado, USA). This instrument combines three spectrometers to cover the portion of the spectrum between 350 and 2500 nm. The instrument has a spectral sampling distance of $\leq$ 1.5 nm for the 350–1000 nm region and 2 nm for the 1000–2500 nm region. The soils were measured using a contact probe (Analytical Spectral Devices, Boulder, Colorado, USA), and a spectralon® panel was used for white referencing once every 10 measurements. For each soil measurement 30 spectra were averaged to improve the signal-to-noise ratio. We collected spectra with a sampling resolution of 1 nm so that each spectrum comprised 2151 wavelengths.

Noisy portions of each spectrum between 350–379 nm and 2451–2500 nm were removed, leaving spectra in the range of 380–2450 nm for our analysis.

A continuum removal technique was used to normalise the reflectance spectra for comparing absorption features from a common baseline. The continuum represents absorptions that are due to different processes than those of interest. It can be calculated using different functions, such as straight-line segments, Gaussian functions, polynomials or splines [13]. In this work, a convex hull was fitted over the top of each reflectance spectrum using straight-line segments that connected local reflectance maxima. For each spectrum, we calculated the continuum-removed (SCR) spectrum (Fig. 2) by dividing the original reflectance values (SR) by the corresponding values of the continuum line (SC).

$$S_{CR} = S_R / S_C \qquad (1)$$

The first and last reflectance values of each spectrum are on the hull, therefore the first and last wavelengths in the output continuum-removed spectrum are equal to 1.

2.3   *Statistical Analysis*

A PCA was performed on the SCR data from which the means were subtracted, (i.e. centered data). We used the iterative NIPALS algorithm [14]. We did not standardize the data to unit variance because all our wavelengths were in

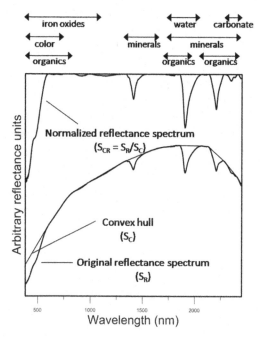

Figure 2. Sample soil vis-NIR spectrum displayed as percent reflectance, convex-hull and continuum removed reflectance. The plot shows regions of the spectrum that hold important information on soil constituent.

the same units and the differences in variability between them were inherently important. The scores condense the information in the samples, and the eigenvectors show the variables (in our case the wavelengths) that load heavily on the particular component. PCA reduces the dimensionality of the data to fewer components that describe a large portion of its variance. The first component accounts for the largest variance, while subsequent components account for decreasingly smaller portions.

The first three principal components where spatialized using geostatistical analysis. Geostatistics was originally used in the mining industries to prospect minerals [15], but has proven to be useful in soil science to describe and understand the spatial distribution of measured soil properties [16]. Geostatistics consists of variography and kriging. Variography uses semi-variograms to characterize and model spatial variance of data, whereas kriging uses the modeled variance to estimate interpolated values between samples.

For the purpose of this study, a semi-variogram was calculated for the first three principal components. The variogram provides a means to quantify the spatial variation for a range of properties (e.g., spectral data) between sampling points separated by a given distance [17]. The model parameters (Fig. 3) nugget, range and sill were determined. The nugget is the positive y-intercept of the model and corresponds to discontinuity of the soil variable, usually arising from errors of measurements or if a sampling interval is too coarse. The range is the separation distance where points are no longer spatially correlated. The sill is the point where the curve levels out and this equals the prior variance of the variable. Experimental variograms were calculated for the first three principal components. A spherical model showed the best fit for the PC1 scores, while a Gaussian model showed the best fit for the PC2 and PC3 scores.

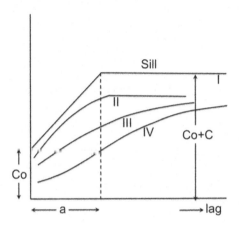

Figure 3. Examples of semi-variograms: I linear; II Spherical; III exponential; IV Gaussian. The model parameters nugget (Co), sill (Co+C) and range (a) are shown.

3   Results and Discussion

Visible-NIR spectra contain information on minerals, organics, water, color and particle size, which are fundamental components of the soil. Here, we will not report a comprehensive account of the important absorption features of vis-NIR spectra, but only mention them in a broad sense and direct the reader to relevant literature (e.g. Ben-Dor *et al.*, 1999 [18]; Clark *et al.*, 1990 [8]; Leone *et al.*, 2000, 2012, 2013 [19–21]; Viscarra Rossel and Behrens, 2010 [9]). Briefly, absorptions in the visible–short wave NIR (400–1000 nm) are due to soil carbon and Fe-oxides (mainly hematite [α–Fe^2O^3] and goethite [α–FeOOH]), while those in the NIR (1000–2500 nm) are due to water, clay minerals, carbonates and organic matter. Figure 2 shows the approximate position of absorptions of some fundamental soil constituents.

The first three principal components condense about 98% of the total variance of the original spectral data (i.e., normalized reflectance spectra). For that, only the scores of these principal components have been retained for further analyses.

The eigenvectors of the first three principal components are shown in Fig. 4.

The eigenvector of the first principal component showed a step increase toward the blue and ultraviolet wavelengths which mainly due to a strong iron-oxygen charge transfer band, associated with the presence of iron-oxides, which extend into the ultraviolet [22]. The higher values of loadings in the visible range of the first principal component, might be partly due to soil organic carbon (e.g., McCauley *et al.* 1993 [23], Shonk *et al.*, 1991 [24]). The eigenvector of the second principal component was dominated by positive loading near 1400 and 1900 nm, which might be due to 1:1 layer clay minerals (mainly smectite), specifically to structural O-H stretching mode in its octahedral layer (1400 nm) and combination vibrations of water bound in the interlayer lattices as hydrated cations and water adsorbed on particle surfaces (1400 and 1900 nm) [25,8]. Finally, the eigenvector of the third principal component had negative loadings near 1400 and 1900 nm (mainly due to smectite, as previously discussed) and near 2200, due to Al-OH bend in the lattice of 1:2 layer clay minerals (mainly kaolinites) [8].

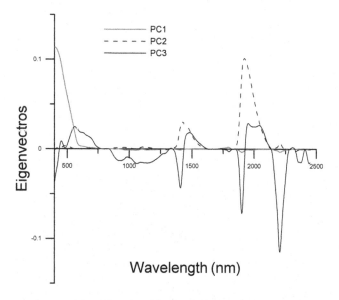

Figure 4. Eigenvectors of the first three principal components (PC1, PC2 and PC3).

Table 1 shows the model parameters of the semi-variograms used for the spatialization of the first three principal components. Figures 5–7 shows the kriging maps of these components.

The PC1-map (Fig. 5), accordingly with the previous discussion, mainly shows the variability of iron-oxides content. Therefore, it is possible to assert that the soils of the southern zone of the investigation field (higher PC1-scores), morphologically more elevated, are characterized, in respect to those of the northern zone, morphologically more depressed, by an higher iron oxides content. This hypothesis is coherent with the geochemical dynamic of iron in soil, which is strongly affected by the redox condition of the medium. Under oxidizing conditions, more frequent in the southern zone, the iron tends to become insoluble, with subsequent formation of iron oxi-hydroxides. Under reducing conditions, caused by a prolonged period of waterlogging, iron composts dissolve rapidly, releasing Fe^{2+}.

Table 1. Variogram model parameters for the first three principal components.

Variable	Model	Nugget	Sill	range	C/C+Co
PC1	Spherical	2510	3022	710	0 17
PC2	Gaussian	1400	2390	590	0.41
PC3	Gaussian	42	105	600	0.60

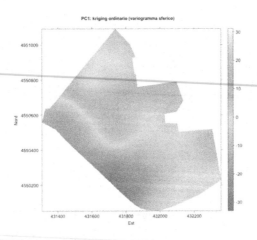

Figure 5. Kriging map for the first principal component.

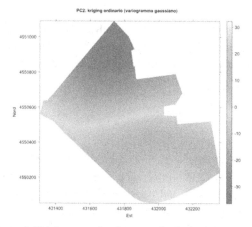

Figure 6. Kriging map for the second principal component.

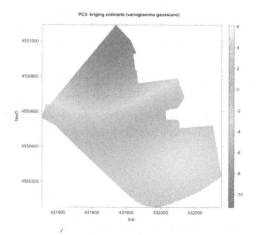

Figure 7. Kriging map for the third principal component.

The PC2-map mainly shows the variability of 1:1 clay minerals (smectite). Therefore, it is reasonable to assert that the northern zone of the investigated field (higher values of scores) has an higher 1:1 clay mineral content. The PC3-map, as already discussed, is instead inversely related to both 1:1 (smectite) and 1:2 (kaolinite) clay mineral contents. This result further confirms the tendency to the increase of the above clay minerals and, consequently, of finer particle sizes in the fine earth, moving from south to north of the field. This statement is coherent with the results of a recent study carried out by Grilli *et al.*, (2014)

[11]. This study, which is based on a traditional soil survey and laboratory analyses, showed that, moving from the south to the northern sites of the field, the clay content (expressed as g kg^{-1}) of top soil increases as follow: 22.0, 31.6, 35.3, 41.4, 43.0. Conversely, the silt content (g kg^{-1}) decreased as follow: 68.2, 56.9, 55.3, 53.7, 49.7.

4   Conclusions

The present paper shows a methodological approach for field-scale soil mapping, based on vis-NIR reflectance spectroscopy, in combination with principal component analysis and geostatistics, and interpret the results in the light of the information content of reflectance spectra.

The results are coherent with those obtained with a previous work based on traditional soil survey and laboratory analyses, thus confirming the usefulness of the presented approach. As proof of this assertion, we can mention the recent introduction of the above approach in some important viticultural systems of southern France (zone of Margaux appellation of the Bordeaux region), where, among other things, the vis-NIR-based digital soil maps were used as essential base layers for the development of computer-based decision support systems. Future researches will carried out to further improve the method and/or to adapt it to the agricultural and pedo-environmental specificity of the investigated areas, in order to optimize as much as possible the results obtained, through the adaptation of agricultural practices to soil spatial variability (precision farming).

References

1) van Vuuren, J.A.J., Meyer J.H., Claassens A.S., 2006. Potential of near infrared reflectance monitoring in precision agriculture. Communications in Soil Science and Plant Analysis, Vol. 37, Issue 15–20, 2171–2184.
2) Delin S., Linden B. 2002. Relationships between net nitrogen mineralization and soil characteristics within an arable field. Acta Agriculturae Scandinavica Section B-Soil and Plant Science, Volume 52, Issue 2.3, 78–85.
3) Delin S., Berglund K., 2005. Management zones classified with respect to drought and waterlogging. Precision Agriculture, Volume 6, Issue 4, 321–340.
4) Bouma J., Stoorvogel J., van Alphen B.J., Booltink H.W.G, 1999. Pedology, precision agriculture, and changing paradigm of agricultural research. Soil Science Society of America Journal, Volume 63, Issue 6, 1763–1768.
5) Odlare, M., Svensson, K., Pell, M. 2005. Near infrared reflectance spectroscopy for assessment of spatial soil variation in an agricultural field Geoderma, **126**, 193–202.

6) Milton E.J., 1987. Principles of field spectroscopy. Int. J. Remote Sens., 12, pp. 1807–1827.

7) Drury S.A., 1993. Image interpretation in geology. Chapman & Hall, London.

8) Clark R.N., King T.V.V., Klejwa M. e Swayze G.A., 1990. High spectral resolution reflectance spectroscopy of minerals. J. Geophys. Res., 95, pp. 12653–12680.

9) Viscarra Rossel, R.A., Rizzo, R., Dematte, J.A.M., Behrens, T., 2010. Spatial modelling of a soil fertility index using vis-NIR spectra and terrain attributes. Soil Science Society of America Journal, 74, 1293–1300.

10) Leone A.P., Bilancia M., Fragnito F., Romano G., Calandrell D., Buondonno A., 2014. Uso della spettroscopia vis-NIR e dell'analisi statistica multivaria a geostatistica per la cartografia della variabilità dei suoli a grande scala. Il caso studuio dell'azienfa GioSole. In "Paesaggi e suoli del Basso Volturno per una frutticoltura innovativa" (A.P. Leone, A. Buondonno e P. Aucelli ed.). Grafica Iuorio, Benevento (In press).

11) Grilli E., Leone A.P., Buondonno A., 2014. I suoli dell'Azienda GioSole. In In "Paesaggi e suoli del Basso Volturno per una frutticoltura innovativa" (A.P. Leone, A. Buondonno e P. Aucelli ed.). Grafica Iuorio, Benevento (In press).

12) FAO. World reference base for soil resources, 2006 World Soil Resources, Reports n.103, Rome.

13) Clark R. N., Roush T. L., 1984. Reflectance spectroscopy: quantitative analysis techniques for remote sensing applications. J.Geophys. Res., No B7, pp. 6329–6340.

14) Martens, H., Næs, T., 1989. Multivariate calibration. Chichester: John Wiley & Sons.

15) Matheron G., 1963. Principles of geostatistics. Econ. Geolo., 58, 1246–1266.

16) Webster R., Oliver M.A., 2007. Geostatistics for environmental scientists. John Wiley & Sons, Chichester, England, pp. 315.

17) Webster R., Oliver M.A., 2001. Geostatistics for Einvironmental Scientists. Wiley, Chichester.

18) Ben-Dor, E. et al , Irons, J.R., Epcma, G., 1999. Soil reflectance. In A. N. Renzc (Ed.), Remote sensing for the earth sciences, vol. 3. pp. 111–1888, New York: Wiley.

19) Leone A.P., Spettrometria e valutazione della riflettanza spettrale dei suoli nel dominio ottico 400–2500 nm. Rivista Italiana di Telerilevamento, n. 19, 1–26.

20) Leone A.P., Leone N., Rampone S.. An Application of vis-NIR reflectance spectroscopy and Artificial Neural Networks to the Prediction of soil Organic Carbon content in Southern Italy. Fresenius Environmental Bulletin, 22, 4b.

21) Leone A.P., Viscarra-Rossel R., Amenta P., Buondonno A.2012. Prediction of soil properties with PLSR and vis-NIR spectroscopy: Application to Mediterranean soils from southern Italy. Current Analytical Chemistry, 8, 283–299.

22) Hunt G.R. (1980) – Electromagnetic radiation: the communications link in remote sensing. In 'Remote sensing in geology' (B.S. Siegal and A.R. Gillespie Eds), Wiley, New York, pp. 5–45.

23) McCauley, J.D., Engel, B.A., Scudder, C.E., Morgan, M.T., Elliot, P.W., 1993. Assessing the spatial variability of organic matter. ASAE Paper No. 93–1555. St. Joseph: American Society of Agricultural Engineers.

24) Shonk, G.A., Gaultney, L.D., Schulze, D.G., Van Scoyoc, G.E., 1991. Spectroscopic ensing of soil organic matter content. Trans ASAE, 34, 1978–1984.

25) Bishop, J.L., Pieters, C.M., and Edwards, J.O. (1994). Infrared spectroscopic analyses on the nature of water in montmorillonite. Clays Clay Minerals 42, 702–716.

MEDITERRANEAN AGENCY FOR REMOTE SENSING AND ENVIRONMENTAL CONTROL: SATELLITE MONITORING AND MAPPING.

Roberto Tartaglia Polcini

Chief Operative Officer, MARSec Spa. - r.tartagliapolcini@marsec.it

Abstract

MARSec (Mediterranean Agency for Remote Sensing and Environmental Control) is a satellite monitoring center which provides satellite imagery and remote sensing based value added services to public bodies and private entities. Data are received, processed, stored and distributed directly by the Company. Data processing is performed by MARSec. MARSec operates at regional, national, and international level. Indeed, receives, processes and distributes near real time remotely sensed data covering the whole Mediterranean area, Central Europe and North Africa, and on board recorded satellite imagery worldwide

1 Introduction

The term 'Earth Observation' (EO) refers to the application of sensing measuring technologies (in situ and remote) to study the Earth's environment and the effects of human activities. This includes measuring technologies and platforms, owned and operated by a variety of entities, from research institutions to public authorities, to international organizations and private commercial interests.

For more than half a century, since the launch of Sputnik I in 1957, Earth Observation satellites have been monitoring our global environment, revealing its fascinating beauty while, at the same time, demonstrating its inherent fragility and exposure to rapidly growing human-induced stresses. Satellites enabled humans to explore the solar system and the rest of the universe, to clearly view many objects and phenomena that are better observed from a space perspective, and to use for human benefit the resources and attributes of the space environment. The unique view from space has given us an improved understanding of the Earth, which is essential to predict, adapt and mitigate the expected global challenges and their impacts on human civilization.

The significance of Earth observations in various sectors worldwide is apparent. The data from EO satellites contribute to sustainable development by providing information, measurements and quantification of natural and artificial phenomena. The synoptic view provided by the satellite imagery offers technologically the most appropriate method for quick and reliable mapping and monitoring of Earth's resources. Change detection through repetitive satellite remote sensing over various temporal and spatial scales provides the most economical means of assessing the environmental impact of the developmental processes, monitoring of bio-species diversity of an ecosystem, and evolution of appropriate plans for sustainable development.

Earth Observation data and derived information are essential inputs to improve human understanding on nature's worth. EO contributes to sound policy decisions in international society by providing scientific information necessary for informed global environmental decision-making and for monitoring our progress on all geographical scales as we explore new development paths aimed at sustainable management of the planet.

2 Mediterranean Agency for Remote Sensing and Environmental Control

MARSec (Mediterranean Agency for Remote Sensing and environmental control) is a satellite monitoring center which aims at providing, mainly to public bodies, products and value-added services obtained from satellite data. Data are received, processed, stored and distributed directly by the Agency. Data processing is performed by MARSec; signals received from satellites are transformed into "products/services", or into data ready to be analyzed or used in various applications by public bodies. MARSec operates at regional, national, and international level. Indeed, remotely sensed data cover the whole Mediterranean area, Northern Europe and North Africa.

MARSec aims to create "Complete Information Systems" able to integrate information and data from different sources and to provide real-time (or near-real time) answers for the decision-making process at political and operational level. Satellite and local remote sensing, indeed, can significantly contribute to the creation of a knowledge-based society, one of the goals that the EU aims to achieve by the end of the decade. MARSec is a point of reference for public bodies in the field of remote sensing, environmental monitoring, and protection from natural and man-made hazards. It also provides tools to have a better knowledge of the territory and to monitor its changes continuously.

Figure 1 – Antenna 4.5 meter in diameter – MARSec, Benevento

Figure 2 – Villa dei Papi, MARSec location in Benevento

2.1 *Recent history*

MARSec was founded as a public agency owned by the Province of Benevento. The executive project was entrusted to the University of Sannio. From the January 2012 MARSec has a new private shareholder, GeoNetSAT, a newCo owned by GeosLab Srl, a company specialized in providing products and services for Public Administrations. GeosLab was founded with the aim of organizing a center of production of digital mapping and development of applications based on GIS (Geographic Information System). Today GeosLab gives high priority to the creation of a center of excellence in Campania in the field of environmental monitoring and control of the territory. At the end of 2010 GeosLab was among the promoters of the establishment of the network enterprise GIS called GeoNetCom. The network sees as partners, major companies with offices in Pisa, Milan, Trento, Rome, Tunis, Geneva, Lugano ... to participate in multiple initiatives. It was also created a new company in Tirana (Albania) and the same is happening in Algeria, to cover the many opportunities for business development in the Balkans and North Africa.

2.2 *Technological equipment*

MARSec has two readout stations that, through its antenna, receive data directly from satellites. MARSec staff manages directly the processing and transforms the signals received into "products" or into data ready to be tested or used in various applications.

Through agreements and contracts, currently MARSec receives data from several satellite owned by some international research centers or space agencies. For environmental monitoring on a large scale, MARSec receives data from NASA satellites - "Terra" and "Aqua" - launched by the United States for studying global changes of the Earth. For this reason, the station is part of the NASA project called Earth Observing System (EOS). This Project put together all the direct readout station in the world that receive NASA satellites.

Figure 3 – Terra satellite, NASA's Earth Observing System

Putting together what every station can see in its footprint, it is possible to have a general idea of the whole globe. The footprint of MARSec (from northern Europe to North Africa, from Portugal to the Middle East), allows to MARSec to be a point of reference in the Mediterranean basin. The sensors on board of these satellites are called MODIS. They can generate different products divided, for convenience, in products for Land monitoring, Ocean monitoring, and Atmosphere monitoring.

Figure 4 – MARSec footprint (the internal polygon is the EPOD footprint)

MARSec receives data also from the constellation of satellites that belong to the NOAA. These were born as meteorological satellites. These are equipped with AVHRR sensors and have five channels with which it is possible to analyze NDVI (vegetation indices) and fires. Even in this case, satellites transmit data for monitoring on a large scale.

For high-resolution images, however, MARSec signed, on August 4th 2005, an agreement with the Israeli society ImageSat International NV, owner of "EROS-A" and "EROS-B" satellites. The agreement grants MARSec to acquire images from these satellites. MARSec is the only Direct Broadcast – in Italy – that receives, directly, EROS high-resolution images for Public Administrations and Defense (from which it is possible to distinguish roads and houses). The "EROS-A" resolution is about 1.8 meters, while that of "EROS-B", launched on April 25th 2006, is approximately 70 cm. EROS products can be used for planning purposes, to control the territory, for urban and agricultural planning, civil protection, cartography updating, etc.

Figure 5 – EROS B, VHR satellite of IMAGESAT INTL

On February 16th 2006, was signed the agreement with the *Canadian Space Agency* for the receipt and use of RADARSAT-1 data. In this case, the sensor on board is a Synthetic Aperture Radar (SAR). SAR works in the microwave region of the electromagnetic spectrum and, therefore, may acquire in any climatic conditions because it is able to penetrate the cloud coverage. With Radarsat-1 data it is possible to obtain products useful - more generally - to users involved in the control of territory and planning of human activities. In particular, through a special procedure (differential interferometry) it is possible to obtained useful products to monitor areas affected by landslides or by active deformation phenomena related to various geological phenomena. Other applications are the detection of oil spill in the sea, the subsidence of buildings and the wind maps. Last year the satellite has completed its mission and, now, is no longer operational.

Since 2009, MARSec is a Ground Receiving Station (GRS) certified by ImageSat INTL® as EPOD Station since 2009. The *Exclusive Pass on Demand* (EPOD) Program enables our Ground Station to autonomously task the EROS satellite and directly receive all the acquired imagery on a selected set passes every year into MARSec commercial footprint (EPOD footprint) established in the business agreement.

MARSec can choose in advance and notify ImageSat INTL® of the relevant orbits for which it would like to have full control. In the selected passes, the MARSec GRS will create and transmit to the satellite the acquisition command file generated by *Mission Planning System* (MPS). The satellite will acquire the images as planned and will downlink them to our Ground Station in real-time. Main advantages are:

- Exclusivity: the GRS has exclusive control over all images acquired on selected orbit
- Autonomy: the GRS has full and independent control to use satellite camera
- Flexibility: the GRS selects imaging parameter (type of images, imaging resolution, scanning angle, time of integration).

As additional capability, the MARSec can ask to the satellite provider, the acquisition of new targets within 2-3 days from the request by CAS order procedure.

Main Areas of Activity

Natural and man-made disasters
- Forestry and forest fires
- Floods and landslides

Urban areas
- Urban planning
- Urban change detection
- Cartography updates

Environmental monitoring
- Coastal zones
- Environmental change detection
- Land use and land cover mapping
- Precision agriculture
- Landfill monitoring

Emergency management
- Emergency Support Systems

3 MARSec best practices in satellite monitoring and mapping

3.1 *Near real time processing of MODIS data for environmental control*

MODIS (Moderate Resolution Imaging Spectroradiometer) is a key instrument aboard the Terra (EOS AM) and Aqua (EOS PM) satellites. Terra's orbit around the Earth is timed so that it passes from north to south across the equator in the morning, while Aqua passes south to north over the equator in the afternoon. Terra MODIS and Aqua MODIS are viewing the entire Earth's surface every 1 to 2 days, acquiring data in 36 spectral bands, or groups of wavelengths. These data will improve our understanding of global dynamics and processes occurring on the land, in the oceans, and in the lower atmosphere.

MODIS is playing a vital role in the development of validated, global, interactive Earth system models able to predict global change accurately enough to assist policy makers in making sound decisions concerning the protection of our environment. Each sensor is able to see Italy twice per day.

MARSec downloads and processes MODIS data in near real time. Data is automatically acquired by an X band antenna when Aqua or Terra satellites are inside station visibility mask (Fig.4) and stored in raw format (since July 2004). Soon after storage, the level 1 processing is started automatically to generate input for the final MODIS products (level2).

Table 1: EOS Satellite pass time on Italy.

satellite	Pass time on Italy	
TERRA	Morning (UTC)	Afternoon (UTC)
AQUA	09:30 – 10:30	20:30 – 21:30

MODIS data are processed daily and automatically at different levels. Level 0 are raw data (measures of radiances and geo-location data on the image, in other words the telemetry of the pass), Level 1a includes digital count for all 36 MODIS bands at all spatial resolutions, detector's views (Earth, solar diffuser, black body, space, spectroradiometric calibration assembly), and information on the data acquisition instruments and on the ancillary data for the aerial platform. These level 1a data are the inputs for the geolocation, the radiometric and geometric calibration and for the following processing. Level 1a also includes data quality indicators to locate missing or bad pixels and particular acquisition mode for the instrument.

Starting from level 1a, the software procedure generates level 1B data (MOD021KM, MOD02HKM, MOD02QKM and MOD03), with this meaning calibrated radiances (W/m2μmsr) geo-located for all 36 MODIS bands extracted from MOD01. This product also includes additional data as quality flags, error valuation and calibration data. Bands 1 and 2 have a 250m spatial resolution (MOD02QKM product), bands from 3 to 7 have a 500m spatial resolution (MOD02HKM product) and the remaining bands have a 1000m spatial resolution (MOD021KM product). Inside MOD03 product there are information on the geodetic coordinates, ground point elevation, satellite and sun zenith angle, and view azimuth angle for each 1km pixel.

Appling empirical and physical algorithms, developed by an international research community, MODIS sensor radiance data are converted level2 products containing bio-physical measurements of the Earth system about atmosphere, sea and land properties.

For example, a level 2 MODIS product regards the localisation of thermal anomalies on test areas. To this end, data are processed taking into account atmospheric variables (that is atmospheric profiles ad columnar water vapour profiles), morphologic variables (land cover and emissivity) and cloud cover (on the test area) to reduce noise effects on TIR signal acquired from MODIS sensor. A bash/python module, at the end of processing (about 20 minutes after the end of satellite passage), generate an e-mail alert in case of thermal anomalies detection.

Every 16 days, are automatically generated level3 products (as NDVI, dNDVI, LST) on Campania region area.

A python module developed at MARSec automatically archive on tape storage level2 and level3 products and generate layer to upload them on a WebGIS site (AIRONE platform) created with open source technologies (LAPP architecture, geoserver/mapserver as geographic server, javascript, php and openlayer as web programming languages).

By MODIS data is possible to have a daily monitoring for long period. This makes it possible to carry out sophisticated analysis on seasonal variables observed, or evaluate the alteration of ecosystems as a result of management decisions or unpredictable events. For this purpose, it is however necessary to have a system for archiving data on-line for quick access and fast processing. The size of this archive was estimated at about 100 Tera Byte data.

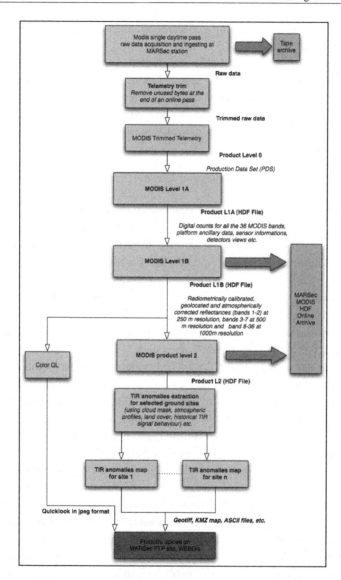

Figure 6: Example of MODIS data processing at MARSec

3.2   *Environmental monitoring and territory control using EROS B data*

MARSec was one of the first Italian companies to think about the use of high resolution satellite images for the monitoring of territorial changes on a large scale. For this reason MARSec has developed the MISTRALS project to address the phenomenon of illegal building, as well as the MEDUSA project, which allows to monitor territorial changes in relation to anthropic and natural hazards.

3.2.1   *MISTRALS: illegal buildings monitoring*

MISTRALS was designed in collaboration with Campania Region, and in particular the Urban Department. The project's objective was to apply innovative technologies as remote sensing, in order to use data acquired from satellites and their processing, to monitor and control the phenomenon of illegal construction, mainly present in regions of South Italy. Remote sensing offers, in fact, an important contribution to the solution of this problem because the acquisition of satellite imagery can be very frequent. Moreover, the high resolution of the acquisitions to determine if there were significant changes in urban areas, can be used to generate immediately an output updated. It reduces the time gap between the urgent need for information and availability of official information reference.

The methodology followed by the MARSEC, about the identification of changes in urban areas, is the result of original algorithms and procedures, developed on the basis of well-established results in the international scientific literature which were later customized by the Satellite Centre in Benevento.

These algorithms and procedures use high resolution satellite images that are processed on a same area in different time. The images processed (georeferenced and orthorectified) and applied on a common reference platform (co-registration) are input for the semi-automatic Change Detection procedure. The combination of automated processing techniques, with the analysis and verification activities carried out by operators, can deliver high standards of quality and accuracy. The data are published on a customized and interactive WEBGIS (http://www.mistrals.it). This solution is very innovative and unique in Italy, for the extension in question (over 13,000 km2 - 2/3 times a year), for the algorithms used and for the modality of distribution of results.

Figure 7: MISTRALS project: Area under control by satellite monitoring

3.2.2   *MEDUSA: monitoring of anthropic and natural risks*

MEDUSA (acronym for Monitoring Environmental Damages Using Satellite Acquisitions) is designed to provide information on territorial changes by combining spatial and temporal information and comparing images acquired at different times in a particular context: the prevention of natural and anthropic risks. MEDUSA works with new or old images (aerial and satellite). The images, orthorectified and co-registered, are submitted to a semi-automatic change detection, which involves four phases: change detection, change quantification, change assessment, and change attribution. The 'change detection' phase is based on the study of the qualitative differences between the areas under investigation. This name is often used to refer also to the whole analysis process. Data exploration to highlight differences can be based on a visual approach, such as when images acquired at different times are overlapped; and it can also be based on the analysis of the variation in the distribution of values for individual pixels in panchromatic or multispectral images. The next step is the 'quantification' of the detected differences, i.e.

change quantification. At this stage, the techniques to be applied must be chosen on the basis of the context in which we operate, taking into account the characteristics of the available data. This is because different and conflicting results can be obtained from the same data. The obtained 'maps of changes' must be further interpreted by determining the significance of differences, i.e. change assessment. The process ends with the change attribution phase, focusing on the identification of the causes of the changes or on the formulation of hypotheses in this regard.

Data are then organized on a WebGIS platform (http://medusa.marsec.it) that allows users to access cartographic databanks related to environmental changes detected. The procedure developed for MEDUSA project is not limited to image analysis for change detection. Indeed, a complete chain is activated, which ends with the reporting of data to the competent authorities and end customers. The WebGIS application was developed using open source technologies in full compliance with industry standards internationally defined by the W3C (World Wide Web Consortium) and OGC (Open Geospatial Consortium). Within the MEDUSA project, MARSEC is developing a new technique to assess volumetric changes so that the third dimension might also be introduced.

3.2.3   *MAPSAT: monitoring of anthropic and natural risks*

The mapSAT© (http://mapsat.marsec.it) service provided by MARSec allows to quickly update technical cartographies and produce new maps, related to portions of land, through the use of EROS B satellite imagery. In both cases, this service aims to update or create 'layers' which can be detected in satellite images (e.g., roads, buildings, infrastructures, etc.) on small and medium scale. The new cartographies are useful for territorial analysis and urban planning.

The technological solution proposed by mapSAT© is based on the use of high resolution panchromatic satellite imagery (0.70 m) of EROS B satellite, useful to quickly update technical cartography on small and medium scale (1:10000 / 1: 5000 / 1:2000). The main innovation introduced by mapSAT© is the possibility to update maps in an "expeditious" and continued way, thanks to "multi-temporal" high-resolution satellite data acquired at MARSEC. The availability of cyclical acquisitions on the same area offers the possibility to continuously control man-made transformations. Thanks to mapSAT©, MARSec is also able to provide a regular update of the cartography in order to have updated maps at any time.

Figure 8: MEDUSA project: WebGIS application

The potential users are public bodies involved in planning activities, positioning of new infrastructures and decision-making processes. Potential users also include companies and public bodies who must comply with the requirements of current legislation and must share the knowledge of the impact of their plans and programs with the stakeholders (e.g., construction companies of renewable energy facilities and large-scale infrastructures).

Acknowledgements

I thank my colleagues that have supported me with their technical contributions, to write this article in which I tried to summarize some of the technical potential and the specific skills that MARSec reached.

For all I remember: G. Meoli and C. Maffei (MODIS), C. Di Paola and C. Mercurio (EPOD), R. Giorgio Gaggia and M. C. Carrabba (MISTRALS), M. Focareta and N. Fiscante (MEDUSA), G. Piacquadio (MAPSAT).

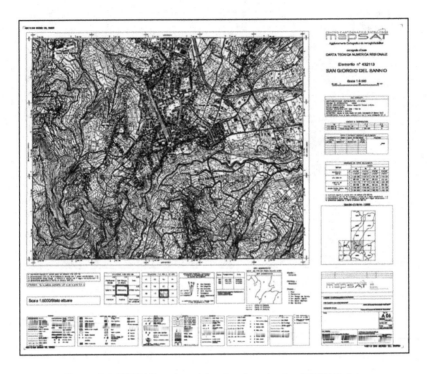

Figure 9: MAPSAT project: Map developed using EROS B data

EVALUATION OF GREENHOUSE GAS EMISSIONS OF E-COMMERCE

Valerio Morfino*
Futuridea, Benevento, Italy
valerio.morfino@ctcgroup.it

Alessandro Perrella
VII Dpt Infectious Disease and Immunology
Hospital D. Cotugno, Naples Italy
alessandroperrella@fastwebnet.it

Salvatore Rampone
University of Sannio
Dept. of Science and Technology
Benevento, Italy
rampone@unisannio.it

Abstract

The use of the Internet has grown continuously in recent years and it is expected it still continues to grow significantly. The eCommerce applications are also growing significantly. In this work we describe an explicit reliable method for calculating the CO_2 emissions of an eCommerce systen, using simple to find data. The methods use variables that are related to hardware and software factors, but also to elements which are not strictly technical such as Web Design and Web Marketing.

---

*Corresponding Author Valerio Morfino (Futuridea).

1   Introduction

Sending an email generates an emission of 4 g of CO_2. This value becomes 50 g if there is a large attachment [1]. A single search on Google produces a greenhouse gas emission of 0.2 g [2]. Facebook produces in a year 13.6 million tons, Skype 24 million tons and whole Internet 300 million tons [3].

The entire Information and Communication Technology (ICT) contributes about 2% of global CO_2 emissions [4-6]. This is the same value produced by the entire avionics industry [7-9].

It is estimated that by 2020, CO_2 emissions from the ICT industry should at least double [6].

The e-commerce applications are growing all over the world in a meaningful way. In the United States the total value of transactions in B2C e-commerce sites has reached, in 2012, 343 billion dollars, an increase of 14% compared to 2011; Western Europe in 2012 has reached EUR 173 billion, an increase of 11% over the previous year. In Italy in 2012 the total value of transactions increased by 18% to almost the total of 11 billion euro. Users of eCommerce websites have grown up by 33% compared to 2011, reaching 12 million [10].

In section 2 we show the method described in the paper [11], in which we have described the components of an ecommerce system, how they contribute to CO_2 emissions and an explicit formulation for the calculation of CO_2 emissions. This formulation included: hardware that hosts the eCommerce website and networking; eCommerce Software; Web Design; Web Marketing; emissions from the shipping.

The formulation excluded the CO_2 emissions generated by the devices used by users to navigate the ecommerce Web Site. Furthermore, the previous formulation, although strict, was not immediate application as needed accurate measurements with specific software or with laboratory instrumentation for evaluating the energy consumption.

In section 3 we describe a simplified version of the formulation that allows an estimate of CO_2 emissions of an eCommerce Web Site with parameters that make use of easy to find values, such as from systems as Google analytics typically connected to the eCommerce sites.

In section 4, with this formula we calculate, as an example, the CO_2 emissions of a small to medium sized e-commerce site. In addition, we show possible improvements resulting from the optimization of Web Design.

2 A method to compute CO2 emissions of an e-commerce system

In this section we summarize the results collected in the paper [11]. An eCommerce website requires several components and services for its operation, such as hardware, software, network connections, as well as logistics services (e.g. shipping). Some of these items – such as freight forwarding – have a direct emission of CO_2, others use directly – like the hardware – or indirectly – such as software – electricity, which is produced, at least for a quota, through the burning of material emitting greenhouse gases.

The energy consumption can be converted into CO_2 emissions using conversion factors typically provided by government agencies [12, 13]. There are also applications that convert kilowatt-hours of electricity in greenhouse gas emissions, such as Greenhouse Gas Equivalencies Calculator provided by EPA - US Environmental Protection Agency [14].

The elements that determine the emission of CO_2 in an ecommerce system are the following:

- hardware that hosts the eCommerce website
- networking
- eCommerce Software
- Web Design
- Web Marketing
- emissions from the shipping
- device used by end user

The total emission of an eCommerce site per unit time is given by the following formulation that includes all items of the list, except devices used by end users:

$$Eec(t) = V(t) \times ((100\text{-}Tct)\ Eecnc + Tct \times Eecc) \qquad (1)$$

Where:

$V(t)$ is the number of visits per time unit. It is related to η_{nd}, efficiency rate of Web Site Surfing related to Web Design work.

Tct is the e-commerce conversion rate per time unit. It is related to η_{wm}, efficiency rate of Web Marketing.

Eecc is the number of total Co2 emission per time unit in the case of navigation with conversion. It is given by:

$$Eecc = Eecnc + L \tag{2}$$

where:

L is CO_2 emissions related to transport

Eecnc is the number of total CO_2 emission per time unit in the case of navigation without conversion. It is given by:

$$Eecnc = Eecp \times Ppv \tag{3}$$

where:

Ppv is the number of page per Visit. This value can be found via Google Analytics [Maltraversi, 2013]

$$Eecp = (Hp + Tgc \times Pcpu + Tgd \times Phd + Tmd \times Phd) \times G \tag{4}$$

where:

Hp is related to power consumption of hardware and Networking

Tgc is the CPU time for the generation of a page. It is related to the quality of the software η_i

Pcpu is the average power output per second of the CPU

Tgd is the Hard Disk time for the generation of a page. It is related to the quality of the software η_i

Phd is the average power output per second of the Hard Disk

Tmd is the average time of the Hard Disk to load multimedia resources of a single web page. It is relate to the quality of multimedia resources η_m

G is the conversion rate from Kwh of energy to CO_2 equivalent emissions

3   Quick estimation of CO2 e-commerce web site emissions

The formulation described in previous excluded the CO_2 emissions generated by the devices used by users to navigate the e-commerce Web Site.

Furthermore, although strict, was not immediate application as needed accurate measurements with specific models [15, 16], software [17, 18] or with laboratory instrumentation for evaluating the energy consumption of the server while ecommerce software run.

For our quick estimate of CO_2 emissions, we assume that for each GB of data processed by an Internet server and transmitted over the network, are used 3.5

kWh of electricity [19-21]. This value changes depending on hardware that hosts the site, the network and the energy performance of the software identified with software energy efficiency rate η_i [11].

The energy consumption of servers and Internet infrastructure is:

$$Egb \times Sp \times Ppv(t) \times V(t) \qquad (5)$$

where:

Egb is the Energy per GB of data

Sp is the Page size in GB

Ppv(t) is the number of pages per visit in the period

$V(t)$ is the number of visits in the period

The energy consumption of customer's devices is:

$$Vd(t) \times Ed(t) \qquad (6)$$

where:

Vd is the average visit duration in hours in the period.

Ed is the average energy consumption of user's device that browse ecommerce web site in the period. This value should consider the various devices used (pc, smartphone, tablet).

So a more simple formula to evaluate total CO_2 emission of an ecommerce Web site is:

$$(Egb \times Sp \times Ppv(t) \times V(t) + V(t) \times Vd(t) \times Ed(t) \,) \times G + Tct \times V(t) \times Lo \qquad (7)$$

where:

Tct is the e-commerce conversion rate per time unit.

Lo is the average value of CO_2 emissions related to shipment of a single order.

G is the conversion rate from Kwh of energy to CO_2 equivalent emissions.

Table 1: How to find parameters values

Variable	Description	How to find Value
Egb	Kwh per Gigabyte of Data	Estimated Constant value 3,5
Sp	Avarage Page Size in Gb	Browser's or online debugging tools
Ppv(t)	Pages per visit in the period	Google Analytics or similar
$V(t)$	Visits in the period	Google Analytics or similar

Vd(t)	Average visit duration	Google Analytics or similar
Ed	Kwh per hour of use	Google Analytics or similar Device Energy Consumption Data
G	conversion rate from Kwh of energy to CO_2 equivalent emissions	Greenhouse Gas Equivalencies Calculator [14] Local provider e.g. [22]
Tct(t)	E-commerce Conversion Rate	Google Analytics or similar
Lo	CO_2 emissions related to shipment of a single order	CO_2 emission online calculator [23]

4   Computation of CO2 emissions of an e-commerce web site

It is now time to compute the CO_2 emission of an e-commerce web site. We use data from a real web site. This test e-commerce site sells all over the world Made in Italy goods. For reasons of privacy we does not indicate the name of the shop and URL.

4.1   *Assumptions*

For simplicity of analysis we overlook the parameters related to the transport. So we do not need Tct(t) and Lo.

The time period of analysis is 1 month, April 2013.

4.2   *Retrieval of experimental data and calculation*

To calculate the average size of the page we used PageSpeed Insights for Google Chrome [24]. The measurement was carried out by selecting 10 pages that represent a typical browsing. The pages have been selected on the basis of the data of "Behaviour Flow" of Google Analytics.

The average page size is 453.99 Kb. The size is inclusive of all linked resources such as javascript, images, css, ecc. Cache using has been considered. Thus, a resource used in more than one page is considered only once.

The value of Page per visit, Visits, Average visit duration are taken from Audience/Oveview section in Google Analytics.

The value of average energy consumption of user's device Ed was calculated considering the time spent browsing through PCs, tablets and smartphones based on Google Analytics (section Audience/Mobile/Overview). To estimate the consumption of the different devices were used manufacturer's data sheets and data from web site http://sympact.cs.bris.ac.uk/webenergy/main/app [25]. The value of G is from [14].

Table 2: Values for calculation

Variable	Description	Value
Egb	Kwh per Gigabyte of Data	3.5
Sp	Average Page Size in Gb	0.0004329
Ppv(t)	Pages per visit April 2013	13.11
V(t)	Visits in the period April 2013	4710
Vd(t)	Average visit duration April 2013	580.72 sec
Ed	Kwh per hour of use	0.04742
G	conversion rate from Kwh of energy to CO_2 equivalent emissions	0.70

The calculation gives the value of 90.68 kg CO_2 per month. About 1090 kg per year.

4.3   Impact of web-design improvement

In Table 3 is shown a simulation in which the parameters refer to the Web Design have an improvement of about 20%.

This kind of improvement can be achieved by:

- improving the navigation, which helps decrease the number of pages per visit and the duration of the navigation (η_{nd} parameter)
- reducing the size of the resources used by the web page (η_m parameter)

Table 3: Values for calculation

Variable	Original Value	Optimized Value
Sp	0.0004329	0.0003463
Ppv(t)	13.11	10.49
Vd(t)	580.72 sec	464.20 sec

The calculation gives the value of 62.7 kg CO_2 per month. About 752 kg per year. About 31% reduction in CO_2 emissions.

5 Conclusions

In this paper we have presented a simplified method for calculating the CO_2 emissions of an ecommerce website. We have shown a case of an ecommerce website for small to medium sized. Finally we have shown that an improvement of Web Design has an important impact on the energy performance and the resulting CO_2 emissions.

Further studies, including other factors in addition to Web-Design factors such as computer and Web Marketing, are required to better elucidate and assess the whole impact of eCommerce in CO_2 emission.

Acknowledgements

The authors wish to thank Alessia Fresilli for her support in experimental data collection and Nicola Scotto di Carlo for precious help in Web Design assessment in this work.

References

1) Lardillon C., Ménard C., Minier N., Communique de Press 7-7-2011 - Courriers électroniques, requête Web, clé USB: quels impacts environnementaux?, Ademe (Agence de l'environnement et de la maitrise de l'énergie), Angers Cedex, France, 2011

2) Hölzle U., Powering a Google search, Google Official Blog http://googleblog.blogspot.it/2009/01/powering-google-search.html, 2009

3) Atkinson A., How Much CO2 is created by..., General Electric Company, http://visualization.geblogs.com/visualization/co2/, 2012

4) Pettey C., Gartner Estimates ICT Industry Accounts for 2% of Global CO2 Emissions, Gartner Press Rel., http://www.gartner.com/it/page.jsp?id=503867, 2007

5) Kilper, D. C.; Atkinson, G.; Korotky, S. K.; Goyal, S.; Vetter, P.; Suvakovic, D.; Blume,O.Power Trends in Communication Networks IEEE J. Sel. Topics Quantum Electron. 2011,17 (2) 275– 284

6) Chan C. A. , Gygax A. F. , Wong E., Leckie C. A., Nirmalathas A., Kilper D. C., Methodologies for Assessing the Use-Phase Power Consumption and Greenhouse Gas Emissions of Telecommunications Network Services. Environmental Science & Technology, 2013, 47 (1), pp 485–492

7) Hanle H., Richter-Shalaby B., White Paper Green ICT, T-systems International GmbH, Frankfurt, Germany, 2010

8) Pettey C., Gartner Estimates ICT Industry Accounts for 2% of Global CO2 Emissions, Gartner Press Rel., http://www.gartner.com/it/page.jsp?id=503867, 2007

9) Gunaratne C., Christensen K., Nordman B., Managing energy consumption costs in desktop PCs and LAN switches with proxying, split TCP connections, and scaling of link speed Int. J. Network Manage. 2005, 15 (5) 297– 310

10) Rangone A., Liscia R., Perego A. Mangiaracina R., Osservatorio eCommerce B2C 2012: I Consumi 2012 in Italia: -2% offline, +18% online... ma la partita è multicanale, School of Management Politecnico di Milano, Dipartimento di Ingegneria Gestionale, 2012

11) Morfino V., Rampone S., Un metodo di valutazione delle emissioni di CO2 di un sito eCommerce B2C

12) CO2 Emissions from Fuel Combustion 2012, International Energy Agency. CO2 Emissions from Fuel Combustion, http://www.iea.org/publications/freepublications/publication/name,4010,en.html, 2012

13) Department for Environment Food and Rural Affairs, 2012 Greenhouse Gas Conversion Factors for Company Reporting,

http://www.defra.gov.uk/publications/2012/05/30/pb13773-2012-ghg-conversion/, 2012

14) http://www.epa.gov/cleanenergy/energy-resources/calculator.html

15) Kansal A., Zhao F., Fine-grained energy profiling for power-aware application design in ACM SIGMETRICS Performance Evaluation Review, Volume 36 Issue 2, September 2008 Pages 26-31ACM New York, NY, USA

16) Koomey J.G., Estimating total power consumption by servers in the U.S.and the world. Technical report, Lawrence Berkeley National Laboratory, USA, 2007

17) SPECpower_ssj2008 benchmark, Standard performance evaluation corporation, http://www.spec.org/power_ssj2008/index.html, 2012

18) Goraczko M., Kansal A., Liu J., Zhao F., Joulemeter: Computational Energy Measurement   and   Optimization,   Microsoft   Research http://research.microsoft.com/en-us/projects/joulemeter/, 2012

19) http://greenalytics.org/data   Zapico, J.L., Turpeinen, M., Brandt, N. (2010) Greenalytics

20) Weber, C. L., Koomey, J. G. and Matthews, H. S. (2010), The Energy and Climate Change Implications of Different Music Delivery Methods. Journal of Industrial Ecology, 14: 754–769. doi: 10.1111/j.1530-9290.2010.00269.x

21) Malmodin, J., Moberg, Å., Lundén, D., Finnveden, G. and Lövehagen, N. (2010), Greenhouse Gas Emissions and Operational Electricity Use in the ICT and Entertainment and Media Sectors. Journal of Industrial Ecology, 14: 770–790. doi: 10.1111/j.1530-9290.2010.00278.x

22) Rapporto mensile sul sistema elettrico consuntivo maggio 2013, Terna Rete Italia, http://www.terna.it/LinkClick.aspx?fileticket=109432, 2013

23) http://www.ecotransit.org/

24) https://developers.google.com/speed/docs/insights/using_chrome

25) http://sympact.cs.bris.ac.uk/webenergy/main/app

PRE-FEASIBILITY STUDY
"SAVE THE CAMELS"

Maria Luisa Varricchio[a], Carmine Nardone[b]
Futuridea, Benevento, Italy
[a]mluisa.var@tiscali.it

Abstract

The feasibility study of the project "Save the Camels" is one of the main goals of the memorandum of understanding between UTAP (Union Tunisienne de l'Agriculture et de la Pêche), GSEEP (Global Sustainable Social Energy Program) and FUTURIDEA (Innovazione Utile e Sostenibile) for promoting technological, scientific and agricultural cooperation. The aim of this project is saving ten thousands dromedaries located in the Kebili's area in central Tunisia.

Although the Camelidae are being able to survive under harsh environment, due to their unique morphological and physiological features, in some areas of Tunisia, they have a nutritional deficit even though they are well adapted to the utilization of vegetation of low nutritional value. Consequently, there are a serious production and reproduction problems, which lead to a decrease in the number of heads.

The camel is an excellent resource, because it transforms vegetal species that cannot be used by humans in very important protein sources for the people living in arid or semi-arid areas. Therefore, it represents a big opportunity for the development of production of milk, meat and cosmetic products; indeed the camel milk also has significant anti-diabetic activity and valuable nutritional properties as it contains a high proportion of antibacterial substances and 30 times higher concentration of vitamin C in comparison with cow milk.

The feasibility study of "Save the Camels" provides a methodological analysis focused on 3 strategic issues supported by the monitoring of the scientific literature.

The working hypothesis are: 1. Localization, spatial distribution and assumptions of food supplements; 2. Nutrition and Feeding; 3. Milk valorisation.

1 Introduction

The genus *Camelus*, which belong to the family Camelidae, includes two species: *Camelus Dromedarius* (generally dromedary), with one-humped, of which the habitat is the dry hot zones of Asia and Africa, and *Camelus Bactrianus* (generally camel), with two-humped, which lives in the cold deserts of Southern areas of the former Soviet Union, Mongolia, East Central Asia and China. From here on, we will be discussed generically call both species as camels.

There are approximately 26 million of camels in the world, of which 91% are dromedaries, however it is difficult to estimate the exact number due to lack of preventive health care services, due to presence of nomadic communities.

In the early 60's, Tunisia had a dramatic reduction of number of camels with about 140,000 – 150,000 heads which increased over the period from 1968 up to reach the 260,000 units. However, in later years, the number has reduced thought the years and the latest data FAOSTAT report the presence of about 237,000 heads in 2012 [1].

In Tunisia, the camels are reared under an extensive system in arid and desert regions to produce meat, milk and fibre. More than 83% of these are found in Tataouine, Kébiliand Médenine Governorates [2,3].

In these regions, the extensive livestock breeding is still traditional and is the most common breeding system among small and medium holders that use the local breeds of sheep, goat and camels and make use, above all, of the natural vegetation without any form of management [4].

The pastoral land is mainly covered with annual grass, acacias, euphorbias and dwarf bushes. The annual rainfall varies between 100 and 400 mm, the amount of rain varying from year to year and the rains being restricted to widely separated areas. This type of pasture permits only extensive types of animal production. Because of its high mobility, its modest fodder requirement and its water regulation perfectly adapted to the environment, the camel is better suited than any other domestic animal to use this type of pasture. According to the nomads, camels can survive in times of extreme need for up to 30 days without water. This depends, however, on the grazing and prevailing temperatures [5].

Camels are slow reproducers; in fact female camel is sexually mature at the age of 4±5 years. Pregnancy is just over 12 months and the calving interval in pastoral production systems is normally 24 months or more. Female camels can remain fertile up to the age of 25 years and, it is often reported, that they produce 8±10 calves during a lifetime. In pastoral production systems, however, only a small proportion of the breeding female can reach this production performance [6].

The new-born calf has no natural protection against diseases, as there is no antibody transfer from the mother during fetal development. The calf can obtain immediate immunization soon after birth only through the colostrum, which has a very high concentration of antibodies.

2 The fundamental role of camels

The decline of the role of the camels as a mean for transport and agricultural work, due to the rapid socio economic changes during the last few decades, has led to the increase in cattle numbers, and hence, a slight decrease in the relative importance of the camel.

Despite from a global perspective, where the economic significance of camel production is minimal in comparison with that of the other domestic animals, in the semi-dry and arid zones, the camels play fundamental socio economic roles and supports the survival of millions of people.

The camel has specific anatomical and digestive characteristics, which facilitate the valorisation of the great arid and desert regions, characterized by scarce and unfavourable resources and to produce milk and meat where other species cannot compete [7]. In fact, they are very reliable milk producers during dry seasons and drought years when milk from cattle sheep and goats is scarce. At such times camel can contribute up to 50% of the nutrient intake of the pastoralists. On the other hand, with growing urbanisation, the demand for milk and animal-source foods among the urban population has been increasing [5].

Therefore, the camels' breeding represent potential opportunities for economic growth and poverty reduction in rural areas.

Camel milk is the only sustenance for the pastoralists for prolonged periods each year; during severe drought periods not only they are able to survive, but continue produce and reproduce while other animals stop the production or die [8].

The large camels (dromedaries and Bactrian camels) are probably the domestic species with the widest range of different functions. Not only do they provide milk, meat and wool (high quality in the case of Bactrian camels), but they also provide energy to transport people and goods and for agricultural activities and are used for leisure purposes, be it racing or rides at tourist sites [9].

Under extreme environmental conditions, such as extreme weather, drought, scarce foodstuff and nutritional resources, the daily milk production is about 2 l/animal [10].

2.1   Dromedary milk traits

It was shown that camels are most important for nutrition of arid and semiarid areas inhabitants. At the present time, because of its special features, there is an increasing interest in camels' milk for human nutrition in other population sectors of different parts of the world [10]. It has been proposed as a substitute for cows' milk in allergic children, as a substitute for mothers' milk for premature new-borns, and as a therapeutic way to repress hyperglycaemia in diabetic patients [11,12]. However, an extensive production system cannot meet the increasing demand or guarantee a constant milk quantity and quality for urban markets. As a result, many intensive dairy camel farms have been recently created around the world, a majority of them using machine milking [13,14,15].

The general composition of camels' milk is different from that of major dairy species and similar in the two specie of Camelidae, camel and dromedary. Dromedary produces diluited milk in hot weather when water is scares. The lactose and protein contents in the milk from the two camel species are similar but their fat contents are different, with camel milk containing more fat than dromedary [16].

The main component of milk, which has a major impact on its nutritional value and technological suitability, is protein. Currently, there are 4 main casein fractions distinguished: $\alpha s1$-, $\alpha s2$-, β-, and κ. Their proportion in the milk is diverse and polymorphism of these proteins was demonstrated in most of the animal species [17-19]. Since that time, many research centers have started exploring polymorphism of various milk proteins in the milk of most of the animals used for dairy purposes.

Perhaps more than any other milk, camel milk has had various therapeutic benefits attributed to it [20]. In fact, camel milk contains greater quantities of bioactive substances and antimicrobial components such as lysozyme, lactoferrins and immunoglobulins than do cow and buffalo milks [21, 22].

The high lysozyme content in camel milk delays growth of yoghurt culture, causing problems in yoghurt production [20, 23]. Even though the antimicrobial components in camel milk are more heat stable than those in cow and buffalo milk, heating camel milk to 100 °C for 30 minutes results in a total loss of antimicrobial activity. [24]. Camel milk may be another good substitute for human milk as it does not contain measurable amounts of β-lactoglobulin, a typical milk protein characteristic of ruminant milk [25]. It is similar to human milk in this respect (Fernandez and Oliver, 1988; Merin et al., 2001; Jirimutu et al., 2010). Therefore, the main whey protein is α-lactalbumin, unlike in cow milk whey in which this protein makes up only 25 percent of the total whey protein. As in human milk, β-casein is the main camel milk casein [20]. These two characteristics could contribute to camel milk having a higher digestibility rate and lower incidence of allergy than cow milk [26]. However, these differences in protein composition are reported to lead to difficulties in cheese manufacture with camel milk [20, 27].

Another crucial anti-allergenic factor is that the functional components of camel milk include immunoglobulin similar to those in human milk, which are known to reduce children's allergic reactions and strengthen their future response to foods [28].

Data concerning the concentration of amino acids in camel milk demonstrate that lysine is the limiting amino acid [22]. Kamal et al. (2007) [30] shown that camel milk contains more methionine, valine, phenylalanine, arginine, and leucine than cow milk.

Fat is the major substance defining milk's energetic value and makes a major contribution to the nutritional properties of milk as well as to its technological suitability. Camel milk, which has the highest state of dispersion of milk fat, contains the most (of the studied animals species) cholesterol (31.3 to 37.1 mg/100 g milk). Camel milk is also unique concerning its fatty acid profile. It contains 6 to 8 times less of the short chain fatty acids compared to milk from cows, goats, sheep, and buffalo [22].

The fat composition is characterized by SFA content (average 60 g/100 g total FA) that may be slightly lower than that of cow milk, while the MUFA content is higher than in cow milk (56 – 80 g vs 26 g/100 g total FAs respective to camel and cow milk) [30]. The main Fa reported in most studies of camel milk are C16:0, C14:0, C18:0, C18:1and C16:1, although a few studies found high contents of C9:0 and C10:1 [31], which are unusual for milk [32,33]. Dromedary milk had a higher proportion of C17:0 and C18:1 than Bactrian milk. The ratio of unsaturated/saturated acid was more favorable in camel's milk compared with that of cows or other mammalians. All of these parameters give a nutritional advantage to camel's milk, although it has a higher content of cholesterol (37.1 mg/100 g) than cow's milk [34]

Milk is also an important source of mineral substances. In particular, camel milk is the richest in iron, zinc, and copper [22].

Camel milk is a kind of exception because of its high concentration of vitamin C [35]. Camel milk contains 30 times more vitamin C than cow milk does, and 6 times more than human milk. This is highly important in desert areas, where fruits and vegetables are scarce. Therefore, camel milk is often the only source of vitamin C in the diet of inhabitants of those regions. [22]. The vitamin C content of camel milk shows a wide range, depending on breed, ranging from 2.5 mg/100 g in the Majaheem breed from Saudi Arabia [36] to 18.4 mg/100 g milk in the Arvana breed from Kazakhstan [37]. However, vitamin C in camel milk may be more heat sensitive than in cow milk, decreasing by about 27 percent when the milk is pasteurized [36].

The camel milk is efficacy on glycemic control risk factors and diabetes quality of life in patients of type 1 diabetes. The action is presumed to be due to presence of insulin/insulin like protein in it. Its therapeutic efficacy may be due to lack of coagulum formation of camel milk in acidic media. Insulin in camel milk possesses special properties that make absorption into circulation easier than insulin from other sources or cause resistance to proteolysis. Camel insulin is encapsulated in nanoparticles (lipid vesicles) that make possible its passage through the stomach and entry into the circulation; some other elements of camel milk make it anti-diabetic (Malik et al., 2012). It has been observed that oral administration of insulin initiated at clinical onset of type 1 diabetes did not prevent the deterioration of beta cell function [38]. Pozzilli et al. (2000) in IMDIAB VII study indicates that addition of 5 mg of oral insulin does not modify the course of the disease in the first year after diagnosis and probably does not statistically affect the humoral immune response against insulin [39]. It is important to note that a certain level of scientific testing on camel milk has already been attempted and documented, particularly, insulin levels in camel milk and this scientific wisdom can be remarkable achievement for diabetic patients [40].

3   "Save the Camels": main purposes

Owing to the increasing human population and declining per capita production of food in Africa, there is an urgent need to develop marginal resources, such as arid land, and optimise their utilisation through appropriate livestock production systems of which camel production is the most suitable without doubt.

In the feasibility study are taken into consideration two fundamental aspects:

- creating and developing a natural oasis of approximately 150/200 km in the Kebili's region (equivalent to that in which the ten thousands dromedaries are currently living in). This purpose allows an optimal management of the existing camels, by improving in a significant way their nutritional and reproductive profiles;
- fostering in connection with the agro-energy project called "Jasmine" the collection and resale of camel's milk for medical purposes, in particular anti-diabetic, even with an economic benefit for the territory (part of whom intended for the maintenance of the dromedaries' natural park).

3.1   *Working hypothesis*

The working hypothesis are subdivided into three projectual ideas, each one of which supported by the monitoring of the scientific literature. The first is the localization, spatial distribution and assumptions of food supplements, articulated in three hypothesis: 1) conservation and enhancement of germplasm of the autochthonous plant species; 2) food supplements both for supporting nutritional deficiencies and improving the nutraceutical quality of milk and meat; 3) satellite tracking (accettable by the local community). The second regarding nutrition and feeding of camels by means of molecular typing of autochthonous vegetal biotypes, selection of autochthonous vegetal biotypes and inventory of real and potential water resources. The third is milk valorization, by means of study of nutritional quality on the basis of relevant information already acquired; definition of strategic guidelines for developing operational solutions; hypothesis of obtaining either new excellence dairy products or cosmetics; identification of nutraceutical properties; genetic improvement program.

References

1) FAOSTAT: Statistics Division [Internet]. Food and Agriculture Organization of the United Nations. Available from: http://faostat.fao.org/. Accessed Mar 20, (2014).

2) M. Hammadi, Caractérisation, modulation nutritionnelle et implication du système IGF dans la fonction de reproduction chez la chamelle (Camelus dromedarius). Ph.D. FUSA GX Belgique, (2003).

3) M. Sghaier, and M. Moslah, Importance du dromadaire en Tunisie et commercialisation de ses produits. Session de formation sur l'élevage camelin, Institut des Régions Arides, Médenine, 18-20 février (2004).

4) M. Chniter, M. Hammadi, T. Khorchani, R. Krit, A. Benwahada, M. Ben Hamouda, Classification of Maghrebi camels (Camelus dromedarius) according to their tribal affiliation and body traits in southern Tunisia, Emir. J. Food Agric. 25 (8): 625-634 (2013).

5) Z. Farah, An introduction to the camel. In: Z. Farah, & A. Fischer (Eds): Milk and meat from the camel, Handbook on products and processing, Vdf Hochschulverlag, Zürich, Switzerland, 15-28 (2004).

6) R. Kamber, Z. Farah, P. Rusch, M. Hassig, Studies on the supply of immunoglobulin G to newborn camel calves (Camelus dromedaries). Journal of Dairy Research, 68 1-7 (2001).

7) V. Laudadio, V. Tufarelli, M. Dario, M. Hammadi, M. Mouldi Seddik, G. M. Lacalandra, C. Dario, A survey of chemical and nutritional characteristics of halophytes plants used by camels in Southern Tunisia, Trop Anim Health Prod 41, 209–215 (2009).

8) M. F. Wardeh e M. Dawa, Camels and dromedaries: General perspectives, in: Current Status of genetic resources, recording and production system in African, Asian and American Camelids (2004).

9) Cirad. www.cirad.fr/en/.../FP4+Faye-camelide-ENG.pdf, (2012).

10) M. Atigui, M. Hammadi, A. Barmat, M. Farhat, T. Khorchani, P.G. Marnet. First description of milk flow traits in Tunisian dairy dromedary camels under an intensive farming system. Journal of Dairy Research, (2014).

11) P. P. Agrawal, S. C. Swami, R. Beniwal, D. K. Kochar, M. S. Sahani, F. C. Tuteja, S. K. Ghouri. Effect of raw camel milk on glycemic control, risk factors and diabetes quality of life in type-1 diabetes: a randomised prospective controlled study, Journal of Camel Practice and Research 10(1): 45-50 (2003).

12) A. Sboui, T. Khorchani, M. Djegham, A. Agrebi, H. Elhatemi, O. Belhadj. Anti-diabetic effect of camel milk in alloxan-induced diabetic dogs: a dose-response experiment. Animal Physiology and Animal Nutrition 94 540–546 (2009).

13) J. Juhasz and P. Nagy. Challenges in the development of a large-scale milking system for dromedary camels. In Proceedings of the WBC/ICAR. Satellite Meeting on Camelid Reproduction, pp. 84–87 (Eds P Nagy, G Huszenicza & J Juhasz). Hungary, Budapest (2008).

14) M. Ayadi, M. Hammadi, T. Khorchani, A. Barmat, M. Atigui, G. Caja. Effects of milking interval and cisternal udder evaluation in Tunisian Maghrebi dairy dromedaries (Camelus dromedarius L.). Journal of Dairy Science 92 1452–1459 (2009).

15) M. Hammadi, M. Atigui, M. Ayadi, A. Barmat, A. Belgacem, G. Khaldi, T. Khorchani. Training period and short time effects of machine milking on milk yield and milk composition in Tunisian Maghrebi camels (Camelus dromedarius). Journal of Camel Practice and Research 17 1–7 (2010).

16) Z. Farah, Camel Milk. In: J. W. Fuquay, P. F. Fox, P. L. H. McSweeney (eds.), Encyclopedia of Dairy Sciences, Second Edition, 3, 512–517. San Diego: Academic Press, Elsevier, (2011).

17) A. Litwi´nczuk, J. Barłowska, J. Kr´ol, Z. Litwi´nczuk. Milk protein polymorphism as markers of production traits in dairy and meat cattle. Vet Med 62:6–10 (2006).

18) J. Barłowska, Z. Litwi´nczuk, M. Kedzierska-Matysek, A. Litwi´nczuk. Polymorphism of caprine milk αs1-casein in relation to performance of four polish goat breeds. Pol J Vet Sci 10:159–64 (2007).

19) J. Barłowska. Nutritional value and technological usability of milk from cows of 7 breeds maintained in Poland. [Post-DSc dissertation]. Lublin, Poland: Agriculture Academy in Lublin. Available from Univ. of Life Sciences in Lublin (2007).

20) O. A. Al Haj, H. A. Al Kanhal. Compositional, technological and nutritional aspects of dromedary camel milk. Int Dairy J 20(12):811–21 (2010).

21) H. El-Hatmi, J. Girardet, J. Gaillard, M. Yahyaoui, H. Attia. Characterisation of whey proteins of camel (Camelus dromedarius) milk and colostrum. Small Rumin Res 70(2–3):267–71 (2007).

22) J. Barłowska, M. Szwajkowska, Z. Litwi´nczuk, J. Krol, Nutritional Value and Technological Suitability of Milk from Various Animal Species Used for Dairy Production, 10. Comprehensive Reviews in Food Science and Food Safety (2011).

23) R. Wijesinha-Bettoni and B. Burlingame. Milk and dairy product in human nutrition. Food and Agriculture Organization of the United Nations, Rome, (2013).

24) E. I. El-Agamy, The challenge of cow milk protein allergy. Small Rumin Res 68(1–2):64–72 (2007).

25) G. Konuspayeva, B. Faye, G. Loiseau. The composition of camel milk: a meta-analysis of the literature data. J Food Compos Anal 22(2):95–101 (2009).

26) E. I. El-Agamy, M. Nawar, S. M. Shamsia, S. Awad, G. F. W. Haenlein. Are camel milk proteins convenient to the nutrition of cow milk allergic children? Small Rumin Res 82(1):1–6 (2009).

27) L. C. Laleye, B. Jobe, A. A. H. Wasesa. Comparative study on heat stability and functionality of camel and bovine milk whey proteins. J Dairy Sci 91(12):4527–34 (2008).

28) Y. Shabo, R. Barzel, M. Margoulis, R. Yagil. Camel milk for food allergies in children. Isr Med Assoc J 7:796–8 (2005).

29) A. M. Kamal, O. A. Salama, K. M. El-Saied. Changes in amino acids profile of camel milk protein during the early lactation. Int J Dairy Sci 2:226–34 (2007).

30) E. Medhammar, R. Wijesinha-Bettoni, B. Stadlmayr, E. Nilsson, U. R. Charrondiere, B. Burlingame. Composition of milk from minor dairy animals and buffalo breeds: a biodiversity perspective. J Sci Food Agric: Nov 14 (Epub ahead of print) (2011).

31) A. M. S. Gorban, O. M. Izzeldin. Study on cholesteryl ester fatty acids in camel and cow milk lipid. Int J Food Sci Technol 34(3):229–34 (1999).

32) E. Muehlhoff, A. Bennett, D. McMahon. Introduction. In: Milk and dairy product in human nutrition. Food and Agriculture Organization of the United Nations Rome (2013).

33) E. Muehlhoff, R. Wijesinha-Bettoni, B. Burlingame, Milk and dairy product composition. In: Milk and dairy products in human nutrition. Food and Agriculture Organization of the United Nations Rome (2013).

34) G. Konuspayeva, E. Lemarie, B. Faye, G. Loiseau, D. Montet. Fatty acid and cholesterol composition of camel's (Camelus bactrianus, Camelus dromedarius and hybrids) milk in Kazakhstan. Dairy Sci Technol 88(3):327–40 (2008).

35) M. S. Y. Haddadin, S. I. Gammoh, R. K. Robinson. Seasonal variations in the chemical composition of camel milk in Jordan. J Dairy Res 75(1):8–12 (2008).

36) M. A. Mohaia. Vitamin C and riboflavin content in camels milk: Effects of heat treatments. Food Chem., 50(2): 153–155 (1994).

37) G. Konuspayeva, B. Faye, G. Loiseau, M. Narmuratova, A. Ivashchenko, A. Meldebekova, S. Davletov. Physiological change in camel milk composition (Camelus dromedarius) 1. Effect of lactation stage. Trop. Anim. Health Prod., 42(3): 495–499 (2010).

38) L. Chaillous, H. Lefevre, C. Thivolet, C. Boitard, N. Lahlou, C. Atlan-Gepner, B. Bouhanick, A. Mogenet, M. Nicolino, J. C. Carel, P. Lecomte, R. Marechaud, P. Bougneres, B. Charbonnel, P. Sai. Oral insulin administration and residual beta-cell function in recent-onset type 1 diabetes: a multicentre randomised controlled trail. Diabetes Insulin Orale group. Lancet 356: 545-549 (2000).

39) P. Pozzilli, D. Pitocco, N. Visalli. No effect of oral insulin on residual beta-cell function in recent-onset Type 1 diabetes (the IMDIAB VII). Diabetologia 43: 1000-1004 (2000).

40) R. P. Agrawal, S. C. Swami, R. Beniwal, D. K. Kochar, R.P. Kothari. Effect of camel milk on glycemic control, risk factors and diabetes quality of life in type 1 diabetes: a randomized prospective controlled study. Journal of Camel Practice and Research 10 45–50 (2003).

A FOOD SAFETY AND TRACEABILITY SYSTEM BASED ON RFID TECHNOLOGIES AND SERVICES

Gianni D'Angelo
University of Sannio
Dept. of Science and Technology
Benevento, Italy
dangelo@unisannio.it

Gianfranco De Luca
Sannio Engineering
Progettazione Integrata
Benevento, Italy
info@sannioengineering.it

Salvatore Rampone
University of Sannio
Dept. of Science and Technology
Benevento, Italy
rampone@unisannio.it

Abstract

Worldwide, foodborne diseases are an important cause of mortality. There is a strong need to strengthen surveillance systems for foodborne diseases. Traceability is an increasingly common element of public and private systems for monitoring compliance with quality, environmental, and other product and process attributes related to food. The key issue to add values on traceability is to integrate the traceability system with the supply-chain management processes, by integrating small-scale farmers-producers in the food supply-chain through ICTs. In this paper, we introduce a traceability system based on RFID technology and Internet of Things.

1 Introduction

Food production and distribution systems are becoming more interdependent, integrated, and globalized. At the same time, the diffusion of foodborne diseases have raised the need to ensure food quality and safety [1]. This need leads to trace food from the manufacturing point to the point of consumption. Traceability is a concept developed in industrial engineering and was originally seen as a tool to ensure the quality of production and products [2]. Economic literature from supply-chain management defines traceability as the information system necessary to provide the history of a product or a process from origin to point of final sale [3]. Food traceability systems allow to all supply-chain actors and to the regulatory authorities to identify the source of a food quality problem and initiate procedures to remedy it. However, some small-scale farmers lack the resources to comply with increasingly strict standards for tracking and monitoring of the supply-chain variables through sophisticated technologies. So, for these farmers, the traceability can represent barriers to commerce. Although, for many food businesses, traceability is seen as a daunting task with few financial benefits, traceability can create competitive advantages. The benefits of traceability for consumers, government authorities, and business operators are widely recognized [4]. The key issue to add values on traceability is to integrate the traceability system with the supply-chain management processes and use the traceability data to manage the business process and improve its performance. Integrating small-scale producers in the food supply-chain through the Information and Communication Technologies (ICTs) appears a valid solution in this challenge. Nowadays, the proliferation and the greater affordability of ICT devices (e.g. of mobile devices) offer potential for small-scale producers to implement traceability systems and connect to global markets. The three main keywords able to improve significantly a traceability system are RFID [5], EPCglobal [6], and Internet of Things [7]. Our contribution is reflected in proposing an information structure for building traceability systems. The remaining of the paper is organized as follows. In section 2, we briefly introduce the RFID technology, EPCglobal and Internet of Things. In section 3, we describe the application of these technologies in the food traceability system, and propose an ICT system. Last section is devoted to the conclusions.

2 Technology overview

2.1 *RFID*

RFID (Radio Frequency Identification) [5] is a technology for the identification and/or data storage. It is based on the storage capacity owned by special electronic devices (named tags) able to respond to interrogation at a distance by the appropriate fixed or portable equipment called readers. A basic RFID system is consisted of three parts: RFID tags, RFID reader and computer processing system. RFID tags are the data carrier of the RFID system. Every tag has a unique code to be attached to the object. Many types of tag RFID exist, but at the highest level, RFID devices are divided into two classes: active and passive. Active tags require a power source either connected to a powered infrastructure or use energy stored in an integrated battery. In the latter case, a tag lifetime is limited by the stored energy.

The Passive RFID is of interest because the tags do not require batteries or maintenance. The tags also have an indefinite operational life and are small enough to fit into a practical adhesive label. A passive tag, for against, is composed by a microchip that contains data storage in a memory (including a universal unique number written in the silicon), an antenna and, a physical support that holds together the chip and the antenna called "substrate" and that can be Mylar, plastic film (PET, PVC, etc.), paper or other materials.

The reader is used to read or write information of the tag by using radio waves. When the reader is near to tag, it emits an electromagnetic field that, through the process of induction, generates in the tag a current that powers the chip. So, the chip can send, through the antenna, all his information to the reader.

The processing system is charged of receiving the tag information transited by readers and processing information.

RFID is generally considered a replacement for bar code. Compared with the bar code, RFID technology has many advantages, including high temperature restraint, larger storage capacity, greater security, encrypted communication.

2.2 *EPCglobal and EPCIS*

EPCglobal [6] is a joint venture between GS1 (formerly known as EAN International) and GS1 US (formerly the Uniform Code Council, Inc.). It is an organization set up to achieve worldwide adoption and standardization of Electronic Product Code (EPC) technology. The main focus of the group currently is to create both a worldwide standard for RFID and the use of the Internet to share data via the EPCglobal Network. Electronic Product Code Information Services (EPCIS) is an EPCglobal standard for sharing EPC related information between trading partners. EPCIS provides important new capabilities to improve efficiency, security, and visibility in the global supply-chain.

2.3   *Internet of Things*

The Internet of Things (IoT) is a novel paradigm that is rapidly gaining ground in the scenario of modern wireless telecommunications. The basic idea of this concept is the pervasive presence around us of a variety of things or objects such as RFID tags, sensors, actuators, mobile phones, and so forth, which, through unique addressing schemes, are able to interact with each other and cooperate with their neighbors to reach common goals [7]. The IoT system architecture is generally divided into three layers [8]: the perception layer, the network layer, and the service layer. Perception layer represents the information origin. All kinds of information of the physical world used in IoT are perceived and collected in this layer, by the technologies of sensors, wireless sensors network (WSN), RFID tags, camera, global position system (GPS), etc. Network layer provides transparent data transmission capability. Using existing communication networks, the information form perception layer can be sent to the upper layer. The service layer transforms information to content and provides application and services for end users, such as logistics and supply, disaster warning, environmental monitoring, agricultural management, production management, and so forth.

3   Food Traceability System

We propose a system based on two main technological benefits that have allowed an explosion of interest in RFID and Internet of Things: the wider availability of very low-cost and higher-range passive RFID tags that require no battery to operate, and the use of the Internet to connect otherwise isolated RFID systems. Aims of the model is twofold: safeguard the image of the manufacturer, distinguishing it from counterfeiters of products and processes; reassure consumer about the quality of the products and implement a system to get food on demand. The proposed model for traceability and food safety is shown in Figure 1.

Figure 1: Proposed model for traceability and food safety. The central server shares data between consumers (little house) and small-scale producers (man with pitchfork).

We use a central server as the pivot of data sharing across consumers and small-scale producers. Central server provide an informative interface to the producers in order to give information on the consumers demand. So the food production is based on effective demand. In this way, the waste production and the time of storage are reduced, that is to say, a money saving. Moreover, the consumer get the food with high quality, into required times and may also query food information of the whole process. This leads to a greater satisfaction of the consumers.

In overall processing of seeding, cultivating and so on, RFID technology together EPCIS service is used to collect data automatically. Information about feed, medicine usage, daily management and detecting is recorded into traceability system server via RFID tags. In the subsequent stages of packing, storage and selling, different tags are attached to the objects to identify the food and its state in the supply-chain. In order to take under control the food quality, it should be essential to record environment information in each stage of the chain. So, using specific sensors (e.g. GPS, WSN), it is possible collects multiple information such as temperature, humidity, intensity of light, transporting information and so on. In particular, in transport management, because the vehicles and goods are all attached with RFID tags, when they pass through some checkpoints, information stored in tags can be captured by the readers. So, using RFID technology can greatly accelerate the speed of delivery, improve the efficiency and accuracy, reduce labor and costs, and ensure accurate inventory control, or even know exactly how many containers are in transit, origin and destination of transshipment, and the expected time of arrival. Thus, each stage of the traditional supply-chain including manufacturers, suppliers, transporters, warehouses, retailers, and the customers, is improved. RFID and Internet of Things ensure that the right goods are available in the right place with no discrepancies and zero errors. It makes the supply chain considerably more precise and improves the efficiency and reliability of the entire chain. As real-time information is made available also administration and planning processes can be significantly improved.

4 Conclusions

Traceability is an increasingly common element of public and private systems for monitoring compliance with quality, environmental, and other product and process attributes related to food. In this paper, a new traceability model, based on the combined use of three different standards (RFID, EPCglobal and Internet of Things), was shown to improve the overall supply-chain. It is shown that using RFID technology and exploiting the potential of Internet of Things, the efficiency of supply-chain management can be improved greatly. On the other hand, the proposed model is able to reassure the consumer about the quality of the products and gives a new system to get food on demand. However, many food businesses are still skeptical about using RFID technology in supply chain management because the traceability is seen as a daunting task with few financial benefits. Fortunately, the proliferation of low-cost devices connected to Internet makes us believe that more and more companies can use RFID technology in their supply-chain management.

References

1) T. G. Karippacheril, L. D. Rios L. Srivastava, "Global Markets, Global Challenges: Improving Food Safety And Traceability While Empowering Smallholders Through ICT - Section 3 — Accessing Markets And Value Chains" in ICT in Agriculture Sourcebook, The World Bank 2011. http://www.ictinagriculture.org/sourcebook/module-12-improving-food-safety-and-traceability

2) B. Wall, "Quality Management at Golden Wonder." Industrial Management and Data Systems, Vol. 94, n.7, pp.24–8, 1994.

3) C.S. Overby, C. P. Wilson, J. Walker, "Retailers Need an RFID Code of Conduct." Forrester Research, 2004.

4) X. Wang, D. Li, "Value Added on Food Traceability: a Supply Chain Management Approach," IEEE International Conference on Service Operations and Logistics, and Informatics, pp. 493-498, 2006

5) G. Borriello, "RFID: Tagging the World," Communication of the ACM, vol. 48, no. 9, pp. 34–37, September 2005.

6) O.G. Sobrinho, C.E. Cugnasca, "An Overview Of The EPCglobal Network," IEEE Latin America Transactions, Vol. 11, n. 4, 2013.

7) L. Atzori, A. Iera,G. Morabito, "The Internet of Things: A survey," International Journal of Computer and Telecommunications Networking, Vol.54, pp. 2787–2805, 2010.

8) X. Jia, Q. Feng, T. Fan, Q. Lei, "RFID Technology and Its Applications in Internet of Things (IOT)," IEEE International Conference on Consumer Electronics, Communications and Networks, Vol., pp. 1282 - 1285, 2012.

WTC (WE TAKE CARE) EXPERIMENTAL SMARTPHONE APP TO FOLLOW-UP AND TAKE CARE OF PATIENTS WITH CHRONIC INFECTIOUS DISEASE: WHICH IMPACT ON PATIENTS LIFE STYLE?

Alessandro Perrella
VII Dpt Infectious Disease and Immunology
Hospital D. Cotugno, Naples Italy
alessandroperrella@fastwebnet.it

Valerio Morfino
Futuridea, Benevento, Italy
valerio.morfino@ctcgroup.it

Abstract

People may experience in their life several types of stressful events that may influence social life. Chronic disease represents a further stress that ay influence not only social life but may also have impact on disease evolution too. Infectious disease represents one of the most important field in Medicine. Currently there is no tool to follow-up patients difficult to treat neither to evaluate emotional and social life of those subjects suffering chronic illness. Here we present our preliminary data on our project WTC (We take Care) a smartphone based App to monitor and follow-up patients with chronic infectious disease. The development of the App actively involved the patients, final users of our App.

1 Introduction

The concept of stress is fundamentally related to organism's adaptation to challenging environmental conditions over time. Studies on the stress have sought to explain mainly two key questions: a) how the body maintains core regulatory functions despite the continual, and often times extreme, perturbations imposed by environmental events, and (*b*) the psychobiological costs and consequences of these dynamic regulatory processes. Indeed stress involved diverse responses that, according to previous researches (Weiner 1992) were orchestrated across several levels of psychobiological functioning, an integrated "whole organism" reaction. In fact while psychological processes happen between the environmental events and the organism's response (Somerfield & McCrae 2000), the stress process expanded to accommodate a host of factors involving individual differences in perceptual processes and cognition. The above mentioned events occur in all individuals, however, stress may strongly increase in other conditions that may change psychological or environment settings as happen during chronic disease. Indeed patients with chronic conditions often have to adjust their aspirations, lifestyle, and employment. Many suffer about their difficulty before adjusting to it. On the contrary others have protracted distress and may develop psychiatric or psychological disorders, most commonly depression or anxiety. A prospective study of general medical admissions found that 13% of men and 17% of women had an affective disorder. The proportion of patients with conditions such as diabetes or rheumatoid arthritis who have an affective disorder is between 20% and 25%. Among patients admitted to the hospital for acute care and among patients with cancer, rates can exceed 30% compared with a prevalence of depression in the community of about 4%-8% (1). These data clearly suggest the need to have a well defined protocols to take care of all these patients. Currently in any outpatients clinic, counseling or wellness test are available to assess patients psychological and health status. However, what we could assess during a visit may not correspond to the rest of everyday life of patients suffering chronic disorders. In this context a new kind of approach is required, to evaluate unnoticeably, constantly but scientifically managed the healthiness and wellness of those patients. In the present time one possible way to monitor a wellness state of a subject may be managed using one of the most used and almost indispensable tool: the smartphone. These devices have such many functions that could help to follow-up and interact with the patients as well as evaluate their physical and psychological status throughout Apps, particular software running on these devices.

Currently, mobile Apps have almost unlimited potential to improve clinical practice, system efficiency as well as to propose easy access to medical update. Up to the present time, Apple store and Google's Play store, taken together, account for more than 13.000 available apps in the Fitness/Medical section (2). In a world quickly increasing its use of mobile technology, this number is likely going to have a rapid spread in the next future.

However there is no clear consensus on how manage the app development, particularly in medical fields, to ensure their safety and usefulness for patients or clinicians, even if FDA proposed selected criteria (3) neither the real involvement of the patients in early phase of development of these apps. Despite these premises, the use of mobile technology in clinical practice is usually based on one way interaction where Hospitals may suggest or remind some treatment strategy or receive or prevent very few information about some behaviours (4). Nowadays the smartphone technology may be used in a more interactive way, where the hospital not only may send an input but it could receive multiple responses from conscious and unconscious reactions of the patients, according to smartphone's sensor to recognize movement, touch (pressure and duration), position and so on (Figure 1). Recently these functions have been separately proposed in several clinical conditions to follow-up and take care of patients (5). However, there are no evidence of some kind of smartphone app using all these functions simultaneously to evaluate healthiness and wellness of the patients with chronic diseases, particularly those with persistent infectious disease as HIV or Viral hepatitis, that it has been reported to have great impact on social and psychological life (5).

2 Methods

In the end of 2012 we designed an app aimed to recognize multiple inputs from patients suffering persistent infection, according to their disease and eventually to undergoing therapeutic schedule. Mainly, this app has the role to assess conscious and unconscious activities of the patients, identifying stress, emotional feelings and evaluating sleep quality (Figure 1) targeting to have a deeper understanding of patients life style according to their clinical conditions. Considering previous report (5) we managed a survey to assess the real impact and appeal of this kind of app for the patients, involving those subjects in the development of the app too. Here we present the results based on the patients evaluation of our app We Take Care (WTC©) and managed on 100 randomly selected patients (age 18-55 years) with chronic infectious disease (HBV 35 pts- HCV 40 pts –HIV 25 pts). Survey was based on 5 simple questions with only two possible answers (Yes or No) (Figure 2).

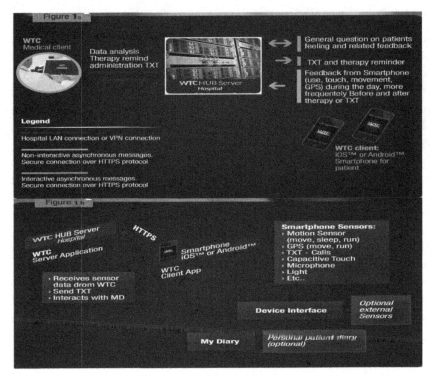

Figure 1: The Figure shows the WTC functioning: It is composed by a client side app (Core and Specialty based Addon) and a server side software (Figure 1a). The app evaluates, during the day and at the time of therapy administration, inputs from smartphone motion sensors (movement, rest), GPS, Capacitive Touch, Microphone (change in voice), Light, TXT and calls (in- and out-coming) (Figure 1b) to assess emotional status conscious and unconscious activities of the patients. Further it can also receive inputs from external devices connected with Bluetooth or NFC. The app send all data to the server that according to patients specific characteristics and based on input received send back Txt or data to better assess patient's behaviours.

Further all data recognized and analysed will be used to give a score (WTC score) that determine the wellness and healthiness status of the subject according to his clinical condition and therapy schedule. This score is the final result of the following simplified formula:

$$WTC_{sc} = sleep_{day} + physical_{day} + social_{day}$$

The obtained result are to add to SF18/SF36 Test acquisition by our App and sensors analysis. Single components of the formula are the results of other equations.

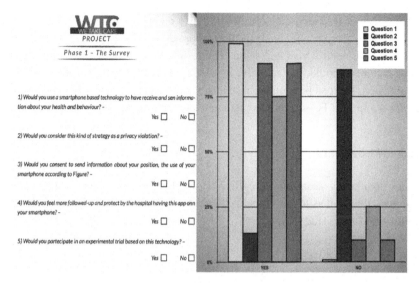

Figure 2: Figure shows results, according to the questions of the survey submitted to the patients. The most relevant results is that highest percentage of the patients in 18-55 years range would use smartphone to evaluate their clinical and emotional status (Question 1) and that they would not consider that as privacy violation (Question 2).

3 Results

Our Study, clearly shows how it would be important to involve patients in the development of any kind of App that is related to disease management. In fact we had that about 95% of all patients would accept to use a smartphone based technology for their disease care. However 87% of the subjects would participate to a trial not considering this strategy as a privacy violation. 90% would consent to send and receive reports, while 75% of these patients would feel more followed-up using this app (Figure 2).

4 Conclusion

In a setting of patients affected by chronic disease, the use of WTC app for smartphone in daily activity, would seem to be appealing for them as well as acceptable with minimal impact on their life style. This is an important results that would strongly propose new and alternative ways to assess wellness and healthiness as well as follow-up of the patients, where the Hospital is not more bonded to its walls and the patients can be examined and followed-up not only on the base of their conscious feeling but also on an unconscious point of view, according to sleep quality analysis, unintentional reaction to predetermined input as therapy or control visit and so on. The results on this survey, strongly suggest that patients should be actively involved in development of this kind of application. Next step would be to manage a clinical trial using this app starting to use a score (WTC score) to be associated to the drug activity characteristics.

Acknowledgments

Many thanks to Costanza Sbreglia MD for her precious help in the patient enrolment as well as to Dr Vincenzo Scarallo for his assistance in psychological evaluation of this study.

References

1) Guthrie E. Emotional disorder in chronic illness: psychotherapeutic interventions. *Br J Psychiatry* 1996; 168:265-273.
2) Buijink AW, Visser BJ, Marshall L. Medical apps for smartphones: lack of evidence undermines quality and safety Evid Based Med. 2013 Jun;18(3):90-2
3) U.S. Food and Drug Administration. Draft Guidance for Industry and Food and Drug Administration Staff—MobileMedical Applications (accessed 20 Aug 2012)
4) Boulos MN, Wheeler S, Tavares C, Jones R. How smartphones are changing the face of mobile and participatory healthcare: an overview, with example from eCAALYX. Biomed Eng Online. 2011;10:24. doi: 10.1186/1475-925X-10-24. http://www.biomedical-engineering-online.com/content /10//24.1475-925X-10-24
5) Muessig KE, Pike EC, Legrand S, Hightow-Weidman LB. Mobile phone applications for the care and prevention of HIV and other sexually transmitted diseases: a review. J Med Internet Res. 2013 Jan 4;15(1).c1

GLOBAL SUSTAINABILITY FOR A WORLD OF 'SMART' BIO-TERRITORIES

Donato Matassino
Former Full professor of Genetic improvement in Animal production
President of ConSDABI –NFP. I. - FAO AnGR
Benevento, Italy
matassinod@consdabi.org

Abstract

The 'sustainability' has to be managed on the basis of innovative pathways taking into account the global biosphere, which can be defined as a complex of 'anthropic-bio-geo-pedo-climatic' factors variable in relation to the geographical area concerned. Hence the need for the 'global sustainability' is the basis for planning human activities in this current millennium. The concept of 'global sustainability' is summarized in the "State of the Planet Declaration" defined in Scientific Conference "Planet under pressure. New knowledge towards solutions" (London, 29 March 2012) and, fundamentally, proposes: (i) reorientation and restructuring of international and national institutions to innovate the governance of the 'Earth system'; (ii) proposal of new objectives of 'global sustainability' taking into account various aspects: food, water, energy safety, biodiversity, ecosystem services, sustainable urbanization, social capital, sea and ocean protection, sustainable consumption and production; (iii) removal of constraints that prevent or reduce to less developed countries to increase their decision-making power in the international dialogue; (iv) market innovation; (v) provision of financial incentives for young entrepreneurs involved in eco-social and environmental initiatives, especially in less developed countries; (vi) promotion of education and scientific interdisciplinary research for a profitable policy orientation towards sustainability; (vii) revision of methods for GDP calculation by introducing indicators that go beyond the purely economic aspects inspired by a 'biology - ecology - economy - culture - social sciences' integrated vision solidifying in 'green economy' or 'bio-economy'.

A fundamental element of the 'global sustainability' is the 'global food security', the theme on which EXPO 2015 "Feeding the Planet, Energy for Life" is focused. In this regard, it would be desirable to optimize the management of what is already produced, such example: (i) planning specific interventions to reduce the discrepancies in the food distribution at the level of planet Earth; (ii) refection and food chain rationalization; (iii) estimate of the food 'nutraceutical' value; (iv) insect farming for an alternative and/or supplementary animal source protein.

The bio-territory, inserted in a novel ecological vision that is inseparable from human context (human ecology), may constitute a 'prototype' of a new approach to manage the kaleidoscopic endogenous resources in order to promote 'global sustainability'. Each 'bio-territory system', defined as "a model of sustainable management of a 'microbioshere' of a given 'geographical area' by local communities", might individuate virtuous routes in relation to its productive potential on the basis of its originality supply. It may be seen as the 'holistic' moment of an 'atomistic' path represented by the multiplicity of its indigenous resources. The bio-territory can be intended to be managed 'smartly' if it firstly identifies its potential and then realize peculiar innovations suited to its real 'productive' potential; everything might be oriented to physical, psychic and social human development, according to the new moral imperative: 'the ethics of care'. Some fundamental prerequisites for a 'smart' management of a 'bio-territory' in line with the 'global sustainability' are: (i) biodiversity; (ii) health geography; (iii) landscape; (iv) culture, research, innovation and training.

It is desirable a new science system of the global environment in which the role of the scientific community is significant in the understanding of 'critical thresholds' of the environmental global crisis in order to harmonize the local interventions in a global context.

1   Introduction

The anthropologic, cultural-philosophical and biological value of the topic we are going to explore today is awesome. Its multidisciplinar and interdisciplinar complexity is matchless and boundless.

The present age suffers of a shaking cultural swing-and-twist and realities haunted by a deep-seated inner poverty and lacks of meaning concealed by overabudance of external apparences. The Socratic 'enlightened ignorance' (Italian: 'dotta ignoranza') stresses nevertheless that the body of knowledge – above of all those about the bios – results from a flown of endless and neverending acquisitions [1].

The boundless variability and complexity result in an abyss of nameable but continually object of search events: continuously changing events, each times different and brand-new in theirs sophisticated flowing, events resulting from an endless combinations of biological and cultural phenomena.

The grandiose and endless multiplicity of cultural and biological phenomena, characterized by a temporal and spatial dynamism in their co-evolution, is holistically able to shape an array of perceptible units that work up their way to the brain. Each human brain has the capability to elaborate this endless flow of information and answer it, but not always in an affirmative (that is: constructive) way. Each person has his/her personal and individual strategy in doing so in which lays its personal epigenetic identity.

Each person, moreover, is able to magnify this stupendous talent of him/her while sailing through the life's chaos, as an organic biological, genetic, epigenetic, brain-led whole, thanks to a deeply-rooted self-awareness. Therefore care and wisdom are utterly required in evaluating skills, ability and clairvoyance of those intended to offer professional services: the human being's biological complexity is almost uncontrollable.

Life's thirst of knowledge induces such a high euphoria and passion that human brain is endlessly requested to reset its hardware in order to respond to the gorgeous variability in information. In doing so the brain yields a continuum, that is a continuous flow in the quest for scientific truth. Furthermore, there's a brain-gender and it lies in the connective mould that is the neuronal framework (brain structural connectome). The "male" brain bears more intra-hemispheric connections and more inter-hemispheric in the cerebellum. The "male" brain, consequently, is more fit for motion activities, that is coordination and speed, and has very marked spatial memory. The "female" brain, thanks to stronger inter-hemispheric connectivity and intra-hemispheric in the cerebellum, is more able to create a sinergy between analytic thinking (originating from the right hemisphere) and intuitive ability (originating from the left): as a result the "female" brain performs better in language, visual memory and social-awareness. That gender-given differences has been mainly detected in individual aged 12 to 14 [2,3].

Culture and science will grow in an integrated unavoidable coevolution that will end eventually in utter integration. People must be inspired not only by hope but trust in their final achievement blending together body and mind, that is going beyond human limits toward real happiness given by pleromic spirit (St. Paul) [4].

Day by day we are confronted with huge, overwhelming waves of documented certainty and scientific evidences. In the temper of the current clash of cultures, brand new human and life rights are standing high. It follows that we need a brand new vision of the Earth sub-system as universe part [1,5].

We assert now that healthy co-evolution's hopes for planet Earth, universe and person itself, lay in each living being's (be it human or not) 'constructivism-ability'[1] [6–11].

The mankind is passing through strong, fast and sudden changes that are setting out fears and hopes together. We are experimenting a transition among daily-events' static vision and a highly fast-moving one: it is fascinating, growing in opportunities and resources but also arising new, profound problems

---

[1] 'Constructivism-ability' expression or 'niche construction' indicates that the evolutive novelthy are a transformation of 'previous potential'; these transformations allow organisms to participate actively to 'construct' the environment in which they live.

in terms of adaptability. We need to be dynamic in elaborating new thoughts and understandings [12].

We are going to synthetically introduce now some of these problems, simply examining the facts. Some semantic elements could help us in pinpoint innovative paths leading to a real enlightened bio-territory management thus to the vision of "smart" bio-territories.

2   Global sustainability

Global sustainability comes from the smart management of bio-territories: or is it just the opposite? It is hard to say.

Sustainability, in productive systems, is [13]:

a) resources-supply: sustainability consists of preservation, regeneration and replacement strategies in order to manage the increasing resources-shortage;

b) operating (functional) integration: in order to achieve sustainability we need dinamic systems for preventing and self-mending human-stress-generated fragilities; operating integration gives plenty of room for multiple-inter-disciplinary approach in: biodiversity importance, time and space; society and ecology relationships.

Global sustainability results from innovative tracks crossing through biosphere managed as constructive global system locally shaped by geographical factors as human impact, biotic impact, geological features, topsoil nature and characteristics, climate. Planet Earth is indeed a real macro-organism, named 'Gaia' the very first time by J.E. Lovelock in 1972: he and L. Margulis in 1974 widened this Gaia theory that depicts planet Earth as a real solid system equitable with an homeostatic entity, able to maintain its dynamic inner-balance and win survivals by itself [14]. Meanwhile in 1785 J.Hutton [15] described the Earth as a self-adjusting super-organism echoing ancient holistic induist concepts. In P.T. De Chardin's opinion (1881-1955) "The biosphere is a developing all-comprehensive complex entity". In V.I. Vernadskij's words (1863-1945) "Biosphere is a gigantic thermodynamic machine linking living and not-living entities".

The Planet Earth could be depicted as an open dynamic, constrained, neghentropic biological system [16–19].

Some of the Earth system's peculiarities lie in sudden changes and no-return points are sometimes generated – on purpose or not – by human activity: a new actual geological era was named after this as "Antropocene" by E. Stoermer [20] first. Hereafter the definition, in P. Crutzen opinion, describes the living eco-system factory now dominated by human activities: it follows that global sustainability must be the cornerstone in planning human activities over this millenium.

Global sustainability vision has been build up during a long way. Sustainability theory was conceived in J. Locke's mind (1632-1704). In the Seventies of the twentieth-century V.R. Potter [21] contributed to it in outlining bioethic doctrine. H.R. Jonas [22] elaborated the doctrine of "taking-charge-of ethics" in order to govern the relationship with the natural environment. In creating the brand new word "bioethics" Potter sharply foresaw how much relevant biosphere-caring is in building social interactions: nature getting along with mankind, nature-human being amity is the only way to reach a fulfilled lease on life not only in the present time but also in the future. It follows that each political program in the matter of enviroment strictly needs such an ethical path [23,24].

Sustainability theory started off by 1972 during first United Nations environment conference in Stockholm as common debating ground on global environmental needs. Here was launched the United Nations Environment program based on both cultural and spiritual values of biodiversity and local communities self-management [25]. Nevertheless the global sustainability theory was in print during 1987 in Brundtland report, commissioned in 1983 and released by world commission on environment and development. J. Boyazoglu [26] rephrasing this report, describes sustainable development as: "development that must fulfill the present-time needs without threatening coming generation's capacities of fulfilling those belonging to them..." According to D. Matassino and A. Cappuccio [23] this is satisfying with three basic requirements:

a) physical sustainability: a resource that belongs to the future and coming generations must be preserved and whole kept unchanged in its reproductive peculiarities;

b) physical-bio-sustainability: the transition from a single resource to an eco-systemic one;

c) physical-bio-social sustainability: globalizing the embedding of the whole relationship-sphere among living beings.

Two basic values are enshrined in point (c), i.e. intra and intergenerational equity.

In an updated vision intergenerational equity should be replaced by an intergenerational improvement model. Sustainability should be granted from generation to generation, always growing stronger. Quoting the philosopher R. Bodei [27]: "Each generation's duty should be to give back more than family and community granted them".

The United Nations conference on environment and development (Rio de Janeiro 1992) stated: "Human beings are the sustainable development focus. Developments right must be guaranteed through:

a) environment defense;
b) worldwide peace;
c) defeating misery.

The World summit on sustainable development (Johannesburg 2002) stated that sustainability strategies must include not only the ecological point of view, but the human person as well. Strategies must be systematic with each human being's right to live in an environment able to enhance his/her whole formation. It follows that sustainable development must be a crucial opportunity "for rebuilding human relationships on the basis of justice and equity".

On occasion of the conference "Planet under pressure. New knowledges towards solutions" (London 2012), the global sustainability idea clearly takes shape in this statement: "The humankind is eating away the whole environment all along its extense, from global to local. Scientific understanding of environment degradation improved in consequence of the meeting in Rio in 1992: but community in reverse was not able to frame environment degradation in the real situation: we should manage the planet as a biophisic complex. Human community's survival, and those both of our civilization and culture depend on: climate stability, natural resouces and eco-systemic services. We have grown as a force of nature but we are individually weak so far. We must reverse our path, we don't have any other option. It's time of act. Let's endorse now the green economy idea that join economic, enviromental and social economy".

The "State of the Planet Declaration" stated some key recommendations:

a) reorientation and restructuring of international and national institutions to innovate the governance of the Earth system;
b) proposal of new objectives of 'global sustainability' taking into account various aspects: food, water, energy safety, biodiversity, ecosystem

services, sustainable urbanization, social capital, sea and ocean protection, sustainable consumption and production;

c) removal of constraints that prevent or reduce to less developed countries to increase their decision-making power in the international dialogue;
d) market innovation;
e) provision of financial incentives for young entrepreneurs involved in eco-social and environmental initiatives, especially in less developed countries;
f) promotion of education and scientific interdisciplinary research for a profitable policy orientation towards sustainability;
g) revision of methods for GDP calculation by introducing indicators incorporating also the 'social equity'.

Some landmarks in rewieving GDP calculation are [28,29]:

a) human development index developed in 1990 by A. Sen (1933-) and Mahbub Ul Haq (1934-1998);
b) Stiglitz-Sen report (2009) issued by CMPEPS (Commission sur la mesure des Performances Économiques et du Progrès Social), established by N. Sarkozy in 2008;
c) the world happiness report, by J.D. Sachs, J.F. Helliwell E R. Layard (2012).

In the perspective of such a review-process, well-being grows more essential than goods consumption.

Particularly, the first Word happiness report proposed Gross National Happines (GNH) as an index of social happiness. The GNH was signed in law in 1972 by king of Buthan and enter into force in 2008. The real social happiness – It stated – doesn't exist if someone else is suffering while others aren't. It exist only in going along in amity both with other human beings and nature. It is possible only thanks an un-conditioned and un-compelled human thinking.

GNH envisions 33 happiness parameters deduced from nine subject matters.

a) psychic well-being (religiousness included);
b) health;
c) timing management;
d) education standard;
e) resilient multiculturalism;
f) accountable government;
g) dynamic human communities;

h) standard of living;
i) resilient bio-diversity.

Sustainability is guaranteed, in D. Matassino opinion [30], only given those binding preconditions:

a) human rights;
b) spreading wealth;
c) dealing justice out;
d) ethic;
e) social accountability;
f) relationship-wise as a prime good;
g) awareness of self;
h) mutual feelings;
i) unselfishness;
j) creative energy.

Those preconditions accordingly impose a new economic theory, based upon a systemic vision dealing with biology, ecology, economy, social sciences and definable as ecologic economy or bioeconomy.

The bioeconomy theory, issued in 1960 by N. Georgescu Roegen, comply with this A. Marshall's [31] statement: "Nature's activity is a complex whole and describing it as a rank of basic statement, in the attempt of simplifing it, results profitless in the course of time". A. Marshall suggests that economy is nothing but a biology branch, broadly speaking. It follows that bioeconomy would be a branch of knowledge in the heart of the life science and no longer matter of profit or benefit [32]. Bioeconomic theory rise up on three pillars [33]:

a) consumer related-biosphere system reorganizing;
b) long-lasting and relationship goods as dynamics improvement of person's well-being tools;
c) natural capital (bio-territory's endogenous resources).

Civic happiness, the same as 'civic economy' [34] or civic behaviour based upon unselfishness and mutual feelings, has to be the ultimate aim of the speculation about how to reach optimal condition on the go through time and space in the frame of an anthropological vision dealing with physical-psychic and civic well-being [33].

The 'economy civic legacy' goes back to Aristotle (384/383 BC) [28,29].

In our opinion 'sharing economy' comply with the needs of alternative economic models suitable for enhancing the global sustainibility. 'Sharing economy' is the propelling agent that press on the border-crossing between private and common goods: on worldwide range the companies are moving towards this model choosing the so-called local-networks. This idea comes from the civic society vision and forerunned thereby the market spirit. We can trail its marks in the 90, through ecclesiastic communities, suburbs and villages where resources-sharing, however small ones, seems to be a suitable model to confront with the storing up ruling spirit. 'Sharing economy' is based upon three cornerstones: gathering, recycling, use frugally. It comes true by [35]:

a) swapping;
b) renting;
c) donation;
d) lending;
e) co-housing;
f) co-working;
g) brainstorming.

A fundamental element of the 'global sustainability' is the 'global food security'. The theme on which EXPO 2015 "Feeding the Planet, Energy for Life" is focused. This theme leads us to consider that:

a) in relation to the 'carrying capacity', estimates on FAO data (www.faostat.org) from 1985 with predictions to the year 2050, at level of planet earth, show the existence of a surplus of availability limited to the total protein destined for food consumption with respect to the need of the population; this surplus, in 2009, amounted to 46 million tons, a quantity that can meet the protein needs of others 2 billion people in addition to the current 7.3 billion; therefore, already considering the production of protein in 2009, there would be a chance to meet the protein requirement of 9 billion estimated for 2050 [36,37];

b) the existence of 'food waste' whose entity amounts:
   (i) at the level of the planet earth, to 1.3 billion tonnes of perfectly edible food; this amount, equal to 30% of the food product, would meet the needs of about 4 billion persons undernourished;
   (ii) at the level of the EU, to about 89 million tons of food scraps (179 kg per capita);
   (iii) at the level of the Italian country, to 76 kg / per capita of food per year.

From what was said it results the need to manage "optimal" what already is produced by making corrective actions and integrative more than enhancing goods productions. Some examples are provided below:

a) rationalization of canteens and distribution chains; hence the need to innovate and improve systems of packaging for the preservation of food with the possibility to consume only small portions;
b) recourse to the breeding of insects ('insect farming') which could provide more sustainable alternative animal protein.

Since the publication of the FAO "Edible Insects - future prospects for the safety of food for humans and feed" ("Edible insects - future prospects for food and feed securty", 2013) will highlight some of the advantages:

a) obtaining high-quality protein, as well as macro and micro-nutrients comparable to those obtained from meat and fish;
b) low risk of transmission of zoonoses;
c) reduced use of water and agropharmaceutical;
d) reduced the problem of management of wastewater;
e) use in the preparation of feed within - today - aquaculture, pig farming and of aviculture.

For example, it is reported that from 10 kg of cereals is possible to obtain:

a) 1 kg of beef;
b) 3 kg of pork meat;
c) 5 kg chicken meat;
d) 9 kg of edible insects.

The achievement of global sustainability requires a regulation due to widespread phenomena such as, for example, the 'land grabbing' (hoarding of the earth / land). Land grabbing is the acquisition by private individuals or by states, of large areas of farmland overseas to produce food for export goods, through contracts for the sale or long term lease.

Since 2000 until today, more than 1,600 agreements were concluded for the acquisition of large extent of land, for a total of over 60 million. Among the top 10 countries 'buyers' are the United States (more than 7 million hectares), Malesia (3 million), United Arab Emirates (2,8 million) and United Kingdom (2,2 million), among the most colonized nations include the Paua New Guinea (about 4 million hectares), Indonesia (3,5 million), Sud Sudan (3,4 million hectares), Democratic Republic of Congo (2, million hectares) [38].

The phenomenon of land grabbing can be as much a chanche than a risk; on the one hand, the acquisitions represent a resource in the economic realities in which these are scarce and necessary; on the other hand, there is a real risk that local people lose their power to control and access the lands sold and the related natural resources. According to Schinaia G. (2014), many of the acquired land remain uncultivated, exploited by monocultures or biofuels; two-thirds of the agreements on the ground in the last 10 years have had the aim of producing raw materials for biofuels, but this is often "destroying the historic relationship between man and land, human habitat and social heritage of the people who lose their land and it's roots".

Therefore, it is crucial to ensure that acquisitions are conducted in order to minimize the risks and maximize opportunities. The factors that favor the phenomenon of land grabbing are:

a) the economic and financial crisis;
b) food emergency;
c) speculation;
d) volatility of agricultural prices on world markets.

The FAO announced the adoption of a code of conduct, through the definition of a framework of international standards to regulate the actions of purchase along lines of transparency and respect for the rights of the most vulnerable [39].

3 'Intelligent' Bio-territory

The management of a 'bio-territory' should result in an irreplaceable contribution to the 'global sustainability'.

Any 'bio-territory can be defined as "a model of sustainable management of a 'microbioshere' of a given 'geographical area' by local communities. (World Resources Institute, World Conservation Union, FAO, UNESCO, United Nations, 1992).

A bio-territory may be seen as the 'holistic' moment of an 'atomistic' path represented by the multiplicity of its indigenous resources [40] (figure 1).

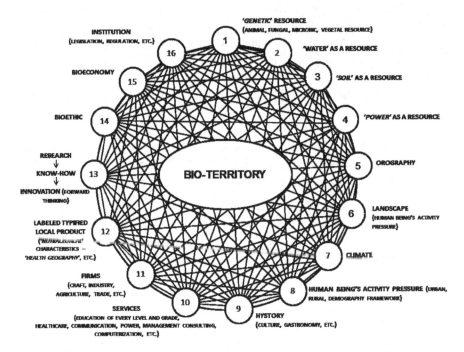

Figure 1. Bioterritory exemplification through the "mandala" [41,12].

A 'bio territorial system' should be analyzed in two ways [42–44]:

a) ecological, including the microbiosphere of a specific geographical area characterized by anthropo-bio-geo-pedo-climatic conditions that differentiates it from others neighboring areas;

b) socio-economic, related to the acting-factors network that regulates the use of local resources and the relationships between the actors and processes of exploitation.

The management of a 'bio territory' requires thinking forward that can be summarized in its intelligent control.

The adjective 'Intelligent' for the development of a 'bio-territory' is mentioned primarily by C. Nardone [45].

The clarification of the word intelligence requires some etymological appeal: the latin term *'intelligere'* (to understand) derives from a contraction of 2 terms: *(a) inter* e *legere* = read 'between'; *(b) intus* e *legere* = read 'inside' [46]. Question:

What does it mean to read 'between' and read 'within' a 'bio-territory'?

Response [42]:

a) read 'between' can mean identifing and understanding the phenomena "anthropo-bio-geo-pedo-climatic" of a bio-territory as a result of relationships between each resource and the area it belongs to;

b) read within may means evaluating every resource endogenous in a perspective of quantity and quality, the use of which must be aimed at: *(i)* inventiveness; (ii) 'innovation' in the 'tradition'; (iii) diversification.

A 'bio-territory' is intelligently operated if it is able to perceive the unpredictable needs and priorities of the system life along the continuous changing in time and space, and to produce solutions that meets the aforementioned requirements. Therefore, a 'bioterritory' is a real dynamic entity in time and space.the bio-territory is 'smart' if it identifies its potential and realizes peculiar innovations suited to its real 'productive' potential; everything might be oriented to physical, psychic and social human development, according to the new moral imperative: 'the ethics of care'[1,46].

According to D. Matassino [40,47], the 'ethics of care' coincides with the well-known moral imperative of the 'alterity' or 'solidarity' that is the "concern for the welfare of another" without necessarily earning back somethings. Paraphrasing G.W.F. Hegel (1770 – 1831) ethics' could be tyhe same as the 'relianza' (report + alliance) of E. Morin [48]; the 'reliance' concentrates in itself that can be identified in ethics: *(a)* of 'love'; *(b)* understanding of the complex or the union; *(c)* of alterity.

The combination of "environmental ethics - personal care" is an indissoluble system since "the environment is the sum of everything that directly affects the probability of living to live and reproduce" then the set of variables: physical, chemical, biotic and psychic influence directly the probability of living to live and reproduce the environment is 'natural' if these variables is added to the man-made environment is considered 'cultural' [12, 49].

According to D. Matassino [12, 50], some of the priorities of management 'intelligent' of a 'bio-territory' can be summarized as follows:

a) revitalization of the economies 'local';
b) inversion of the outputs of resources;
c) block the destruction of the 'bio-resource' local (animal, fungal, microbial, plant) in order to maintain the high 'genetic load' and 'genetic variation';
d) modification of current models of 'production' and 'consumption' in order to reduce their contribution to the deterioration of the environment;
e) change of life styles which constitute 'risk factors' for the security of a 'cultural' agri-ecosystem;
f) restoration of 'campaign - city' integration;
g) rational urbanization;
h) sustainable management of land;
i) rational use of energy from renewable sources;
j) assumption of responsibility for a cultural change by: *(i)* school; *(ii)* the organs of communication; *(iii)* how they function drivers of change;
k) profound attention to 'human ecology ', understood as the "Study of the relationship that human groups have with the various natural and human ecosystems in order to meet their needs with a view to achieve the greatest possible independence, given the resources available in the ecosystem " (C. Raffestin, Centre for Human Ecology in Geneva, 2000); the 'ecology of man 'is an inseparable relationship between ' natural ecology ',' human ecology 'and ' social ecology.

The focal point of planning in management 'intelligent' of bio territory will be the identification – the least wrong possible – a strategy based on a mutual harmonious relationship between 'environmental protection' and 'socio-economic development'. Pursuing this track it is possible to find solutions that reconcile the needs of the two ecosystems 'natural' and 'cultural', in order to get closer to the biological time of 'young' ecosystem, characterized by a positive energy balance: net production always > 0 [12].

The 'intelligent management' of a 'bio territory' the characterization of the geographical area of relevance should be the basis for any programming management. This characterization provides for the treatment and management of a multiplicity of data identifiable with 'big data'.

The 'big data' represent a large and complex collection of datasets. The need to provide a collection of datasets is linked to the need to simplify the management, and then analyze jointly a mass of data, extracting more information than that which could be obtained by analyzing the individual components; 'four v' of big data help us to understand and describe what each of them is [51]:

a) volume = ability to capture, store and access to large volumes of data;
b) speed = ability to perform real-time analysis or almost ('sometimes two minutes it is already too late');
c) variety = refers to the various types of data from different sources;
d) accuracy = quality of data intended as informational value that is able to extract.

As part of a 'bio- territory', the collection of 'big data', varying in time, it can allow you to focus on innovative management models designed to introduce management methods that optimize the 'sustainability' of the 'bio-territory' concerned.

3.1 Some evidence for management 'intelligent' of a bio- territory

Question:
What are some 'prerequisites' for a 'smart' management of a 'bio-territory' in line with 'global sustainability'?

Answer:
The response is rather articulated and complex, since, as mentioned earlier, the bio-territory is a 'system' dynamic optimization given by the contribute of many interacting factors; therefore, it would be impossible to assign a ranking of priorities. Now, only the following aspects will be considered for their international relevance: (i) biodiversity; (ii) health geography; (iii) landscape; (iv) research and innovation.

3.1.1 Biodiversity

A fundamental prodrome for a 'smart' management of a 'bio-territory' is the biodiversity to be considered as a real 'identity card' of a given 'bio-territory' as an expression of a continuum of interaction between 'cultural evolution' and 'biological evolution'.
Biodiversity can be defined as:

a) "every type of variability among living organisms, including, among others, terrestrial, marine and other aquatic ecosystems and the ecological complexes of which they are part; this includes diversity within species, between species and of ecosystems"(Conference in Rio de Janeiro,1992);

b) "'variability' of life and its processes including all forms of life, from single cell to more complex organisms, in all processes, paths and cycles that link living organisms, populations, ecosystems and landscapes" (DG AGRI, 1999).

The 'biodiversity' can be considered [52,53]:

a) a real "stone of wisdom" on which to build a future more and more tended to elevate the lives of humans and other living beings and, in the broadest sense, the life of the planet Earth;

b) real light of a memory founding;

c) "the germ of the life", especially the 'old' and 'local' or 'indigenous' or 'endemismic' one.

The management of the natural resource to be inserted in the concept of 'social capital', to be understood as the set of rules and relations that allow to operate on mutual trust in the prosecution of efficient collective interests [54].

The harmonious relationship between 'cultural evolution' and 'biological evolution' suggests the utility in the management 'intelligent' of a bio territory, of an 'ecosystem approach' takes the form in an evolution 'integral' and 'integrated' of services related to ecosystem functions of 'natural capital' [14,55]:

a) services of "supply" (food of animal and plant origin, precursors of drugs, etc.);

b) service of "regulation" (climate regulation, etc.);

c) services of "support" (pollination, dissemination of seeds, water management, disease control, soil protection, etc.);

d) services "psycho-cultural" (scientific discoveries, use of free time, etc.);
e) services of "conservation" (protection and management of native or endogenous genetic resource).

From 'nutraceutical diversification' deriving from biodiversity, 'nutritional destinations' for 'person' in relation to sex and, within this, can be implemented [36, 37,56]:

a) demographic category (infant, boy / girl, teenager, adult, over sixty, over eighty, centenarian);
b) physiological status (pregnancy, lactation, sports practices, etc..).

The management of 'urban' is showing interesting 'openings' and 'chinks' in a review of 'landscape ecology'. The results of detailed analysis 'ecological' show that biodiversity 'urban' is still full of roots 'native' and 'precious' own of a 'local' and 'old native' diversity derived from species 'endemic' that populate micro-green areas and parks; for example, in the city of Rome, about 115 species of birds recorded, some 106 (~ 92%) are 'native' and only 9 (~ 8%) are 'exotic'. These species 'endemic' reflect, in fact, just the biological uniqueness of the 'microbiosfera' of a specific 'geographical area'. Therefore, urbanization is a harbinger of a loss of 'biodiversity', but it can also play a vital role in the protection of species 'native'. This reality can be attributed, primarily, to the 'ability to constructivism' of any living organism. This new vision 'ecological' of the 'urban' is, presumably, a real change of 'paradigm' so the city should be considered, in this case, not only reduction factor of 'biodiversity', but also a possible tool for protection of species 'endemic' at risk of extinction, representative of the biological diversity in a specific 'region'. Among the species to be protected include those belonging to chiropters for their role ('insectivores'), in urban areas, in integrated control against mosquitoes and sandflies [50,57].

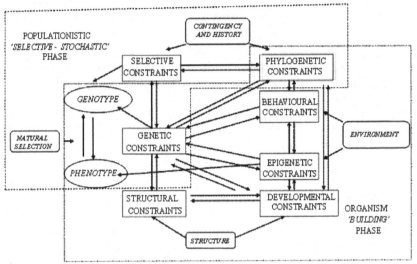

Figure 2: Network of 'constraints' and its probable 'operating system' [60; modified by D. Matassino et al. (61)].

3.1.2 Geography of Health

A 'second' prodome is the concept of "geography of health" [58,59] which occurs through a cycle of mutual interaction can be expressed as follows [29]:

a) during embryo development is acquired, in physiological conditions, an optimal and unique metabolism;

b) this optimal metabolism is only possible with a personalized nutrition in terms of nutrigenomics and nutriepigenomics.

Holistically, the living entity to be considered as a 'biological whole' the construction of which is influenced by a complex of 'constraints' (figure 2).

The 'geography of health' is a true scientific revolution, a new paradigm epistemological and hermeneutic [29]. The concept of "geography of health" makes it ever more indispensable the trinomial 'person - diet - bio territory' ending, in [33]:

a) development of 'urban agriculture';

b) restoration of reciprocity and integration "countryside-town".

An important aspect of integration "countryside-town" is represented by the 'urban gardens' or 'social gardens' in order to obtain 'food' of high-value 'nutraceutical' related to 'geography of health' in the sense that they are peculiar to the 'microbiosfera' in a specific 'geographical area'. The 'urban gardens', once synonymous of poverty or emergency during the war in the 40s, today they are considered an expression of modernity and research of a new 'lifestyle' takes the form in [29,62,63]:

a) consumption of food produced 'locally' and according to the 'rhythm of the seasons'; such food can represent real food derived from 'biological 'agriculture; these foods 'local' are very different from food derived from agriculture so-called 'biological', which are often produced thousands of kilometers away from the place of residence of the consumer;

b) recovery of relationships and activities' social;

c) support to medical therapies (horticultural therapy = therapy with gardens) with type benefits:

   (i) behavioral (reduction of stress, hardship, etc.. due to the direct contact with nature);

   (ii) cognitive (increased self-esteem linked to satisfaction in watching a plant grow and to use the fruits);

   (iii) physical (increase in motor activity, stimulation of visual tactile and olfactory capacity,).

The garden, as long as realized in areas 'unpolluted' ('prerequisite' fundamental) can be considered an example of 'degrowth' conception of political, economic and social that proposing a model of development based on:

a) reduction of consumption;

b) autoproduction;

c) consumption of goods;

because the idea that the 'growth' driven by developed economies to produce more and for all the positive effects in the long term is now put seriously in discussion. As growth 'infinite' in a world of resources 'finite' is by definition 'unsustainable', development 'sustainable' can be considered a true 'oxymoron' from which the antinomy, richer in terms of semantic of 'abundance frugal' [32,64–67].

The vision of 'geography of health' has been fully integrated with the concept of 'mediterranean diet' where the term 'diet' is to be interpreted, however, in its original etymological meaning (probably Hippocratic origin):

from the latin '*diaeta*' that, turn, derives from the greek $\delta \iota \alpha \iota \tau \alpha$ = lifestyle '. In this meaning the 'diet' includes, among other [43]:

a) 'ingested food';
b) 'physical activity';
c) 'socio-economic relations' between people.

The importance of the 'mediterranean diet' is not represented both by the specificity of food and nutrients they contain, as in the 'methods' used to characterize and analyze it and in the philosophy of 'sustainability' that constitutes its essence. In fact, most of the 'values' expressed by the so-called 'mediterranean diet' is part of the following definition of 'sustainable diets'[2]: "'Diets' to 'low environmental impact' that contribute to the 'safety' 'food and nutraceutical' and a 'healthy living' for present and future generations contribute to the 'protection' and 'respect' of 'biodiversity' and 'ecosystem', are culturally acceptable, economically equitable and accessible, adequate, safe and healthy in terms nutraceutical and simultaneously optimize the natural and human resources." In the context of the definition of 'sustainable diet', the so-called 'mediterranean diet', which is strongly associated with the 'microbiosfera of a specific geographical area', represents fully the example of 'sustainable diet', not only for its 'special foods' and 'nutrients', but especially for the 'sustainable lifestyle' that it expresses [68, 69, 70].

Some 'essential points' on which it would be appropriate to base the 'mediterranean diet' are [71]:

a) cultural element;
b) lifestyle pursued by the populations present in the geographical areas is 'coastal' and 'non-coastal' facing the Mediterranean basin;
c) time-sharing (family, community, etc.);
d) element of strong connection to the local community with the bio territory;
e) symbolic element of the universe religious and local traditions;
f) element for value nutraceutical.

In support of the importance of considering the 'diet' as 'lifestyle' show some results of elaboration, carried out by ConSDABI, of data concerning the epidemiological study 'EPIC' (European Prospective Investigation into Cancer

---

[2] Definition coined in occasion of International Scientific Symposium "Biodiversity and sustainable diets united against hunger" [Roma (sede FAO), 3 – 5. XI. 2010].

and Nutrition). The study was intended to investigate, on 520,000 subjects from 10 European countries, the relationship between 'diet', 'lifestyle', 'environmental factors' and' incidence of colorectalcancer (CRC) [28] (table 1).

Table 1: Total meat consumption (g / day), indexed value of the meat consumption ('total', 'red', 'processed' and 'white') making of 100 that of Spain and incidence of colon and rectum cancer (CRC) (ASR), distinctly for European country (including the relative size of the sample) participating in the EPIC project and for geographic area [28; reworking on EPIC data project (72)].

EUROPEAN COUNTRY	PARTICIPANT NUMBER	CONSUMPTION					CRC INCIDENCE	
		TOTAL (g/die)	INDEXED VALUE MEAT					
			TOTAL (%)	READ (%)	PROCESSED (%)	WHITE (%)	ASR	INDEXED
SPAIN	41,438	135	100	100	100	100	11.3	100
NETHERLANDS	40,072	125	93	95	134	45	14.4	127
GERMANY	53,088	120	89	73	151	45	15.7	139
DENMARK	57,053	115	85	102	95	50	19.2	170
ITALY	47,749	113	84	89	63	97	10.9	9
UNITED KINGDOM	87,932	90	67	59	73	71	12.4	110
GREECE	28,561	63	47	64	20	50	8	7
MEDITERRANEAN COUNTRIES (WEIGHTED MEAN VALUE)	117,748	109	80	87	66	87	10	91
MIDDLE-WEST EUROPEAN COUNTRIES	238,145	109	80	78	106	56	15	134

Table 1 shows the following considerations [7,28]:

a) Compared to the spanish, the danes, while consuming about 15% less meat, they have a chance of developing colorectal cancer (CRC) by 70% more;

b) The 'Mediterranean countries', while consuming the same amount of average daily meat (109 g) of the countries at the 'Europa Centre-West', have a chance of developing colorectal cancer by 43% less;

c) Meat consumption reported by the official statistics are to be considered as "apparent consumption" because in the estimate of per capita consumption is considered basically the dead weight of the animal including bone and adipose tissue as well as other non-edible, excluding only 'skin' and 'guts';

d) The 'lifestyle' rather than the individual food consumption, represents a parameter 'basic' to be taken into consideration in studies aimed at defining the 'possible' association 'between' nutrition' and 'incidence of cancer.

In the light of the definition of 'diet' as 'lifestyle', you agree to the following statement of V.A. Sironi [73] "The 'food' as nutrition of 'body', the 'culture' as the food of 'mind' and 'spirituality' as nutrition of the 'soul' are the cornerstones of a dynamic interactive dimension can realize food strategies consolidated and aware, able to combine harmoniously 'taste' and 'pleasure' with 'well-being', 'health' and 'sustainability' in a global perspective".

3.1.3 Landscape

The 'landscape' is an expression of identity of a 'bio territory' because result of the interaction between 'environment and cultural uniqueness of the human community who work there'; therefore, it should be considered in view of an essentially dynamic since the evolution of production techniques influence the manner in which humans change the natural environment [43].

The following definitions summarize very well the emblem of the 'landscape' in context 'bioterritorial':

a) "The form that man, in the course and in the interests of its agricultural production, consciously and systematically, imprints to the natural landscape" [74];

b) "An agro-territorial system inclusive and / or supplemented of / from forms of the natural landscape (soil, water, climate, landscapes, natural resources, plant and animal biodiversity, biocapacity, etc..) of / by the landscape 'cultural' (painting, photography , poetry, prose, music, etc..) of / from the man historically performed by humans in rural areas (production systems, labor, technology, rural architecture, etc..) of / by the 'perception' of men and women " [75].

The 'landscape' plays an irreplaceable role in promoting the 'physical, mental and social human welfare' due to its effect 'restorative' on the psyche of 'person'; hence the necessity of [76–78]:

a) recovery of green areas;

b) restoration and enhancement of aquatic components (rivers, lakes, ponds, etc.).because scientific evidence demonstrating the substantial power 'restorative' of 'blue' color;

c) realization of artefacts and different architectural constructions and chromatically variegated.

The 'polychrome' should be interpreted as 'the absence of homologation'. The 'symbolical' of the 'color' is an essential aspect in peoples' lives (chrome anthropology); J.W. Goethe, in 1810, elaborated the theory of colors (farbentheorie). Today, there are a 'chrome diagnostics' and a 'color therapy' ('calming' colors vs 'exciting' colors and 'exciting' vs 'depressing') which they refer both to fundamentals of psychoanalysis and a know 'traditional'; the latter reflects a plurality of systems and hierarchies of symbolic, liturgical, heraldic and emblematic and alchemical. In the world 'Akkadian-Assyrian-Babylonian' the towers, known as 'ziqqurat', are built on seven floors, with reference to the 7 planets traditionally known and each of which are associated with a metal and a color [50, 79, 80, 81]:

a) gold - the sun;
b) silver - moon;
c) copper - green - Venus;
d) iron - red - Mars;
e) etc.

The effect 'restorative' of 'polychrome' would be associated with the 'beauty' understood as ability to evoke 'awe' or 'wonder' in the observer. The 'surprise' leads, inevitably, to 'knowledge'. H.U. Balthasar (1905-1988) says: "... it is not the 'beauty' that has abandoned us, we who we are no longer able to see it ". In a 'liquid' modern society [82] there has been the disappearance of the 'surprise'. G. Péguy (1873-1914) shows that the disappearance of the 'surprise' is due to a look 'used'. B. Forte [83] emphasizes that the disappearance of the 'surprise' and the ability to discriminate the 'beauty' calls for the need to "open up to a new-found 'philocalia'[3]" [50,83,84]:

The 'beauty' [83]:

a) evokes, not capture;
b) invokes, does not pretend;
c) causes, not sated.

The restorative effect of nature can be obtained by taking care of the 'perspective' of architecture; therefore, even a landscape 'urban', if it is able to induce attention 'not focused', but 'common', may be 'relaxed'; this effect can be obtained through [50,78]:

a) the 'architectural richness' understood as the number of different elements distinguishable into a building;

---

[3] From the greek: φιλο- (the theme φιλέω = love) + καλὸς (= beauty) = love for beauty.

b) diversity among isolates - a group of buildings all the same, though very ornate tends to give a sense of uniformity, while a 'variety' of simple buildings increases the perception of 'complexity'.

3.1.4 *Research and Innovation*

In a 'futuristic vision' of a management 'intelligent' of a 'bio territory', science is responsible to be continually founding of the 'new'; the 'new' one is to be transferred operationally for the constructive evolution of the whole system [46].

Limited to agriculture, some examples of innovations in 'technical' and 'biotech' aimed to a management 'intelligent' of a 'bio territory' able to incorporate production systems, social and cultural existing, not substitute for them, are [85]:

a) characterization of the soil, with particular reference to its micro component;

b) use of models to predict short-term and long-term of 'scenarios' of sustainability for homogeneous areas in line with the definition of a 'bio territory';

c) utilization of alternative energy (photovoltaic, mini - wind plant, etc.);

d) irrigation planned through:

   (i)   implementation of systems that maximize the efficiency of water use while ensuring the prevention of risks of soil salinization in arid areas;

   (ii) rationalization of water accumulation;

   (iii) reuse of treated waste water as a means to produce non-conventional water resources;

e) use of 'satellite informative' for the management of livestock and cultivations;

f) detection with drones;

g) precision agriculture, which takes into account the actual needs of the animal reared and crops;

h) use of so-called 'green architecture' based on the use, as a building material, of plant fiber (hemp, coconut, cotton, wood, straw, cork, etc.).; this material, also referred to as 'renewable matrix', often comes from sources on site; therefore, it contributes to an 'integral and integrated development ' of a given 'bio territory' (for example, construction of shelters for animals);

i) use of 'innovative biotech' latest generation tend to intensify in plants the production of biomolecules 'functional' to human welfare;

j) advanced computerization (for example, , 'anthropization of the computer') - computer 'self-conscious' characterized by a solidarity between components that become capable of exchanging energy according to the need of operation);

k) promotion of so-called 'sector of knowledge' through technical assistance, in order to inform and train workers about the evolution of the environmental risks;

l) incentive of a new information system uniform at national level ("open data") in order to facilitate :

    (i) the selection of animals and plants with regard to specific genetic variants for new 'characters', such as disease resistance, thermotolerance, the' feed efficiency, docility and the obtaining of products' nutraceuticals', using marker assisted selection (MAS), gene assisted selection (GAS), and the genomic selection, especially with the use of panels of SNP (single nucleotide polymorphisms = single nucleotide polymorphism);

    (ii) development of innovative indicators relative to animal welfare and greenhouse gas emissions.

In the context of innovation, microbial plays a fundamental role; just think that [43,86]:

a) as regards the human microbiome, only the gut microbiome to known today amounts to ~ 3.3 million DNA segments encoding the polypeptide ('genes') (from ~ 1,000 species) vs only 20,115 'genes' present in the human genome;

b) as regards the metagenome of the soil, the microbial genome per gram of soil amounts to well ~ 1012 (1.000 billion) base pairs.

The microbiome of the planet earth constitutes one of the most interesting in that it participates directly or indirectly, to sophisticated phenomena such as 'psycho-physical' balance of the person and the production of energy. The human microbiome can be considered a 'system epigenetic' complex and infinite can provide vital contributions to the essential physiological processes such as: (a) digestion (metabolic regulation); (b) growth; (c) self-defense (immune system regulation); (d) psycho-physical balance. In the mouse, there are numerous evidences for the existence of bidirectional communication between 'intestinal microbiota' and 'brain' mediated nerve 'vague', which acts as a 'transmitter-receiver' of signals from the 'intestinal microbiota' to the 'brain' and conversely. The metagenome 'soil', in addition, contributes to: (a) genesis and structuring of the soil; (b) productivity; (c) the health of cultivated plants [28,43].

Microorganisms are very promising for the production of energy. The metabolic activity of microbial:

a) influence the flow of matter and energy through the biosphere;
b) driving cycles bio-geo-chemical global.

The microorganisms, in fact, have the ability to use substrates soluble chemically different for the production of energy. Recently, it has been shown a metabolic versatility of some microorganisms, which utilize acceptor / electron donors in the solid phase through a process called 'extracellular electron transfer (EET)'. In addition to the ecological advantages obtained by the EET system, there is a substantial interest in [87]:

a) take advantage of the microorganisms 'photoautotrophic' for the generation of energy and biofuels;
b) identify 'photoautotrophic' genetically tractable able to use electrical current as electron donors;

Both are prerequisites for a future application in electrosynthesis.

4   Conclusions

1.   We need a new 'system of science of the global environment. This system should be the role of the scientific community in the understanding of the "critical threshold" of the environmental crisis 'global' in order to harmonize the interventions 'local' in context 'global'.

2.   The management of a bio territory "intelligent" according to the canons of its 'global sustainability' implies the need for 'homo oeconomicus' turns into 'homo bioeconomicus' [42].

3.   An innovative development of a bio territory "intelligent" must be in harmony with the context of the 'ecology of man' seen as a ratio of 'alterity'.

4. The bio territory may constitute a 'prototype' of a new way to manage the kaleidoscopic endogenous resources present in line with the logic of the 'global sustainability'.

5. The main problem with the start of this new millennium will be to reconstitute a certain 'encyclopedia' of knowledge in order to give concrete serious and disinterested answers [1].

6. It can be considered that the only force of reason stagger towards the whole of the team of cultural, political, educational and ethical problems [1].

7. The vortical advances in biology, and especially their operational significance, given the important social, ethical, economic and environmental, leave 'disoriented' not only the so-called 'men of the road', but also the researchers and scientists pressing the public to an open debate about the meaning and value of knowledge and the advancement of science for the physical, mental and social welfare of 'person' [88].

8. Science must always be integrated in the social context and, as such, is subject to all the imperfections and limitations that characterize every social activity; as a result of this, it is the right of the 'consumer' to discuss issues related to scientific discoveries [88].

9. The problem of 'ethics' is gaining more importance in the society because of the speed with which humanity today tends to a total globalization of human activity, but with a strong emphasis of a company to 'civilization multiple' so multicultural, multiethnic and multietica [12].

10. The 'pluralism' is a great behavioral philosophy, as it must be understanding of others, it must be a moral and behavioral question and should not be a matter of faith and intellectual ability [1,12].

11. Paraphrasing F. Tessitore (2009): "It is worrying that the pluralism 'ethics' and 'epistemological' has been considered a serious risk, the crisis be overcome and that relativism, in the strong sense of relativity and rationality of knowledge, has been confused with the 'ethical

indifference'; both are not only two serious errors, but two forms of ignorance of the present".

12. Pontiff J. Ratzinger, in chapter IV of his encyclical 'Caritas in Veritate' (2009) in paragraph 48 connects the theme of 'development of peoples' even to 'the duties arising from our relationship with the natural environment'. "The natural environment is considered to be God's gift to everyone, and his use is a responsibility for us towards the poor, towards future generations and towards humanity. .... also the nature is a 'vocation'. Nature is at our disposal as a gift of the Creator who has designed its intrinsic order that the man to draw from it the principles needed to 'guard' and 'nurture'. But it must also be emphasized that it is contrary to authentic development to view nature as more important than the human person... Moreover, we must also reject the opposite position, which aims at total technical dominion, because "the natural environment is not only matter to be disposed at our own pace..." as it goes "... a 'grammar' which indicates ends and criteria for its wise use, not its instrumental and arbitrary". Today much damage to the development originate from these distorted notions. Reducing nature to a set of simple facts ends up being a source of violence to the environment and even encouraging activity that fails to respect to the very nature of man. The human beings interpret and shape the natural environment through culture, which in turn is given direction by the responsible freedom, attentive to the dictates of the moral law. The plans for an 'integral human development' can not ignore coming generations, but need to be marked by solidarity and inter-generational justice, taking into account a variety of contexts: ecological, juridical, economic, political and cultural",

13. As reported in the aforementioned Encyclical recalls the following statement of St. Augustine: "Man must qualify ('frui') of the immense resource that nature puts at his disposal, but should not be used ('uti') this wealth only for itself' [12].

14. Science and technology can not automatically ensure neither progress nor stability on planet earth. They can provide the tools that human race operates successfully in order to achieve dynamic targets useful for the achievement of physical psychic 'person' regardless of his culture and his ethnicity social (D. Matassino, 2010).

15. Paraphrasing P. Mantegazza [90]: "Living is for everyone, the good life is short, live with science, knowledge and conscience is very few."

References

1) D. Matassino, Etica e biodiversità, in Atti VI Convegno Nazionale "Biodiversità: opportunità di sviluppo sostenibile", (ed. Tecnomack), Vol. I, 27 (Bari, 2001).

2) M. Ingalhalikar et al., PNAS, 111 (2) 823, (2014).

3) D. Matassino, Dal rispetto della biodiversità alla medicina di genere, Tavola Rotonda "La Medicina di genere",Benevento, 14 marzo 2014, http://aspa.unitus.it/matassino/1_elenco_pubblicazioni_Matassino.pdf.

4) D. Matassino, La zootecnia in un parco, in: Atti Conv. 'Il parco come punto d'incontro di problematiche socio-economiche di un territorio, con particolare riferimento alla zootecnia', [Tignale (BS), 6 giugno 1997], 9 (1997).

5) D. Matassino et al., Biodiversità e filiere produttive zootecniche, in: Atti 7. Convegno Nazionale Biodiversità 'L'agrobiodiversità per la qualificazione delle filiere produttive', Italus Hortus, 13 (2), 70 (2006).

6) D. Matassino, Lezioni Corso di Miglioramento genetico degli animali in produzione zootecnica, Facoltà di Agraria (Portici), Università degli studi di Napoli 'Federico II' (1975).

7) D. Matassino et al., Vegetarianismo: unica scelta possibile per una corretta nutrizione?, in: Atti Tavola Rotonda "Bioetica e vegetarianismo", nell'ambito del 2. Meeting Internazionale di Bioetica della Biosfera - Ambientamente 2, in press; Il Picentino, XLVIII, gennaio-aprile, 3 (2014); http://aspa.unitus.it/matassino/1_elenco_pubblicazioni_Matassino.pdf.

8) R. C. Lewontin, Theoretical population genetics in the evolutionary synthesis, in: E. Mayr e W. Provine (Eds.) 'The Evolutionary Synthesis', 58 (ed. Harvard University Press), (Cambridge & London, 1980).

9) R. C. Lewontin, Il sogno del genoma umano e le altre illusioni della scienza, (ed. Laterza) (Bari – Roma, 2004).

10) F. J. Odling-Smee et al., Niche construction: the neglected process in evolution, (Princeton University Press), (Princeton, 2003).

11) F. Morganti, Recensione del volume "Niche construction: the neglected process in evolution", Eds. F.J. Odling-Smee, K.N. Laland and M.W. Feldman, (Princeton University Press), (Princeton, 2003).

12) D. Matassino, L'Allevatore, 48 (17), 18 (1992), http://aspa.unitus.it/matassino/1_elenco_pubblicazioni_Matassino.pdf..

13) P. B. Thompson e A. Nardone, Livestock Production Science, 61, 111 (1999).

14) D. Matassino, Biodiversità animale di interesse zootecnico, Documento per il Comitato Nazionale per la Biosicurezza, le Biotecnologie e le Scienze della Vita (CNBBSV), 7 gennaio 2008.

15) J. Hutton, Abstract of a Dissertation read in the Royal Society of Edimburgh, Upon the Seventh of March, and Fourth of April , MDCCLXXXV, Concerning the System of the Earth, its duration and stability, Edimburgh, (1785).

16) L. Von Bertalanffy, Die Naturwissenschafter, 28 (33), 521 (1940).

17) T. M. Bettini, Produzione Animale, 9, 229 (1970).

18) D. Matassino, Il miglioramento genetico degli animali in produzione zootecnica, Eserc. Accad. Agr. di Pesaro, Serie III, 9, 33 (1978).

19) D. Matassino, Problematiche del miglioramento genetico nei bovini, in: Atti XIX Simp. Int. di Zootecnia "Nuove frontiere della selezione per gli animali in produzione zootecnica", (Ed. Clesav), 11, (Milano, 1984).

20) P. Crutzen, Benvenuti nell'Antropocene. L'uomo ha cambiato il clima, la Terra entra in una nuova era, (Ed. Mondadori), (2005).

21) V. R. Potter, Perspective in Biology and Medicine, 14 (1), 127 (1970).

22) H. Jonas, Das Prinzip Verantwortung, Insel Verlag, Frankfurt am Main (1979) [Trad. it. Il principio di responsabilità, (Einaudi), (Torino, 1990)].

23) D. Matassino e A. Cappuccio, Costs of animal products and standard of living, in: Proc. of 8th World Conference on Animal Production, Special Symposium & Plenary Sessions, 559 (Seoul National University), (Seoul, Korea, 1998).

24) E. Sgreccia e D. Tortoreto, Medicina e morale, 53 (5), 887 (2003).

25) D. A. Posey, Cultural and Spiritual Values of Biodiversity, (UNEP) (Nairobi, Kenya, 1999).

26) J. Boyazoglu, Sustainable Agriculture, Animal Production Development and the Environment, in Symp. on Livestock and the Environment, (Korean Society), (Seoul, 1992).

27) R. Bodei, Generazioni, (Laterza) (Roma, Bari, 2014).

28) D. Matassino, Tutela della biodiversità e salute umana, Convegno "Tradizione alimentare dell'Appennino Campano e prevenzione dei tumori", [Acerno (SA) 6 ottobre   2012], http://aspa.unitus.it/matassino/1_elenco_pubblicazioni_Matassino.pdf..

29) D. Matassino, Italiaetica, 7, 14 (2014).

30) D. Matassino, Bioterritorio intelligente in funzione della geografia della salute, Convegno "Modernizzazione e sviluppo del sistema agro-pastorale in Capitanata: dall'indagine storica alla realtà attuale", (Foggia, 15 novembre 2012), http://aspa.unitus.it/matassino/1_elenco_pubblicazioni_Matassino.pdf.

31) A. Marshall, Principles of economics, (Ed. MacMillan), (London, 1890) [trad. it. Principi di economia, (Utet) (Torino, 1953)].

32) D. Matassino, Ricerca avanzata, politiche agro-alimentari e sviluppo territoriale, Convegno "La scuola va nel mondo del lavoro", Morcone (BN), 24 settembre 2009, http://aspa.unitus.it/matassino/1_elenco_pubblicazioni_Matassino.pdf.

33) D. Matassino, Il Picentino XLVI (numero speciale), 26 (2011).

34) A. Genovesi, Delle lezioni di commercio, o sia d'economia civile, da leggersi nella cattedra intieriana, dell'abate Genovesi, regio cattedratico, Fratelli Simone Napoli, 2

voll. (parte prima, pel primo semestre: 1765; parte seconda, pel secondo semestre: 1767).

35) D. Motta, Economia condivisa non solo una moda, Noi Genitori (supplemento Avvenire, Anno XVII, 87) (2014).

36) D. Matassino, Il miglioramento genetico nei bovini per la produzione di latti finalizzati all'uomo, in: Atti Conv. 'Il ruolo del latte nell'alimentazione dell'uomo', Paestum, 24÷26 ottobre 1991, Quaderni Frisona, maggio 1992.

37) D. Matassino et al., Management of consumption, demand, supply and exchanges, in: Proc. Symp. 'On the eve of the 3rd millennium, the European challenge for animal production', EAAP n. 48, (Wageningen Academic Publishers), 105 (The Netherlands, 1991).

38) G. Schinaia, La corsa alla terra dei poveri: accaparramento "selvaggio", Avvenire, 28 agosto, A03 (2014).

39) L. Cotula et al., Land grab or development opportunity? (Ed. IIED; FAO, IFAD), (2009).

40) D. Matassino, L'etica dell'ambiente prodromo della cura della persona, Incontro-Dibattito "Alimentazione e ambiente tra valori etici e tradizione", Benevento 21 febbraio   2014, http://aspa.unitus.it/matassino/1_elenco_pubblicazioni_Matassino.pdf.

41) D. Matassino, Attività zootecniche, in: G. Zucchi 'L'Agricoltura nelle aree protette: vincoli ed opportunità', (Ed. Accademia Nazionale dell'Agricoltura – Ministero per le Politiche Agricole e Forestali), 209, (Bologna, 2005).

42) D. Matassino e M. Occidente, Italiaetica, 7, 8 (2011).

43) D. Matassino et al., Alcune riflessioni conclusive, in: Atti IX Convegno Nazionale sulla Biodiversità, [Valenzano (Bari), 6-7 settembre 2012], in press, http://aspa.unitus.it/matassino/1_elenco_pubblicazioni_Matassino.pdf..

44) W. Nunziatini, Progetto VAGAL (Valorizzazione dei genotipi animali autoctoni), in: 'I Georgofili, Anno 2012, Serie VIII- Vol. 9 (188. dall'inizio), Tomo II/2, 1009, (Firenze, 2013).

45) C. Nardone, Crisi e Sostenibilità, Giornata della Innovazione "Territorio: idee, progetti, prototipi per un nuovo sviluppo sostenibile" - 14. Settimana della Cultura Scientifica e della Creatività Studentesca , Benevento, 20 aprile 2012 (2012).

46) D. Matassino , Un approccio innovativo per la valorizzazione di un bioterritorio 'intelligente', Giornata dell'Innovazione in Agricoltura - idee progetti prototipi per un nuovo sviluppo sostenibile, Benevento (MUSA), 15 giugno 2012, http://aspa.unitus.it/matassino/1_elenco_pubblicazioni_Matassino.pdf..

47) D. Matassino, "Etica e complessità del biosistema", Corso di alta formazione per dirigenti e professionisti "Etica, diritto, economia e cura della persona: ruoli manageriali e attenzione alla persona nella società civile", (Ateneo Pontificio 'Regina   Apostolorum',   24   maggio   2013), http://aspa.unitus.it/matassino/1_elenco_pubblicazioni_Matassino.pdf..

48) E. Morin, La Méthode T I V, L'Ethique, Ed du Seuil, (2004).

49) D. A. Maelzer, J. Theor. Biol., 8, 141 (1965).

50) D. Matassino, Bio-territorio intelligente, Settimana dell'Innovazione in Alta Irpinia - "Smart Rurality", Calitri (AV), 21 marzo, http://aspa.unitus.it/matassino/1_elenco_pubblicazioni_Matassino.pdf.

51) S. Rampone et al., Advanced Dynamic Modeling of Economic and Social Systems Studies in Computational Intelligence, 448, 79, (2013).

52) D. Matassino et al., ARS, 126, 30 (2010).

53) D. Matassino, Convegno "La dieta mediterranea nella prevenzione dei tumori della mammella, del colon - retto e della prostata, Comune di Pollica, Pioppi (SA), 28 settembre 2013 http://aspa.unitus.it/matassino/1_elenco_pubblicazioni_Matassino.pdf..

54) C. Nardone, Cibo Biotecnologico. Globalizzazione e rischio di sviluppo agro-alimentare insostenibile, Ed. Hevelius (Benevento, 1997).

55) D. Matassino, L'importanza del recupero di tipi genetici autoctoni, in: Atti II Congresso Nazionale RIRAB (Rete Italiana per la Ricerca in Agricoltura Biologica) - IX Convegno ZooBioDi (Associazione Italiana di Zootecnia Biologica e Biodinamica) "Il contributo dell'Agricoltura Biologica ai nuovi indirizzi di politica agro-ambientale: il ruolo della ricerca e dell'innovazione", Roma, 13 giugno 2014, I Quaderni ZooBioDi "La Biodiversità: una risorsa per la zootecnia biologica", 9, 17 (ZooBioDi, 2014).

56) D. Matassino, L'apporto della ricerca genetica alla soluzione dei problemi inerenti al miglioramento quanti-qualitativo della produzione del latte e dei suoi derivati, in: Atti 1° Meeting sugli aspetti igienici e legislativi del latte e del gelato, Camaiore (LU), 29-30 aprile, 1983, Nuova rassegna di legislazione, dottrina e giurisprudenza, 60 (16), 1854, Ed. Noccioli, (Firenze, 1986).

57) M. F. J. Aronson et al., Proc. R. Soc. B, 281, 20133330 (2014).

58) M. Hanson, Salute globale: un approccio evoluzionistico, Sigma Tau - XXIII Spoletoscienza "Geografie della salute", Spoleto 3 luglio (2011).

59) M. Hanson et al., Progress in Biophysics and Molecular Biology, 106, 272 (2011).

60) M. Sarà, L'evoluzione costruttiva, Ed. UTET, (Torino 2005).

61) D. Matassino et al., Genomica e proteomica funzionali, in: atti Convegno "Acquisizioni della Genetica e prospettive della selezione animale", (Firenze, 27 gennaio 2006), I Georgofili – Quaderni 2006–I, 201, (Società Editrice Fiorentina, 2007).

62) S. Bertini, Oggi è uno stile di vita, Il Divulgatore 7-8 "Orti Urbani" (2012).

63) G. Gianquinto e F. Orsini, In origine il bisogno di cibo, Il Divulgatore 7-8 "Orti Urbani" (2012).

64) A. Nardone e D. Matassino, Large-scale operations with special reference to dairy cattle, in: Proc. Int. Symp. on the contrainst and possibilities of ruminant production in the dry subtropics, (eds. E. S. E. Galal et al., Cairo, Egypt, 5-7 novembre 1988), EAAP Pubbl. n. 38, (Wageningen Academic Publishers), 167 (The Netherlands, 1989).

65) S. Latouche, Come sopravvivere allo sviluppo. Dalla decolonizzazione dell'immaginario economico alla costruzione di una società alternativa, (Bollati Boringhieri) (Torino, 2005).

66) S. Latouche, Come si esce dalla società dei consumi. Corsi e percorsi della decrescita, (Bollati Boringhieri) (Torino, 2011).

67) S. Latouche, Per una decrescita frugale. Malintesi e controversie sulla decrescita, (Bollati Boringhieri) (Torino, 2012).

68) R. Capone, Risorse naturali e diete sostenibili, in: Atti IX Convegno Nazionale sulla Biodiversità [eds. G. Calabrese et al., Valenzano, Bari, 6-7 settembre 2012], vol. II "Biodiversità, Alimenti e salute", XVI (CIHEAM-IAMB, 2013).

69) C. Scaffidi, Cibo, territorio, identità: la valorizzazione sostenibile, in: Atti IX Convegno Nazionale sulla Biodiversità . Biodiversità [eds. G. Calabrese et al., Valenzano, Bari, 6-7 settembre 2012], vol. II "Biodiversità, Alimenti e salute", XIV (CIHEAM-IAMB, 2013).

70) D. Matassino, Le carni: caratteristiche organolettiche e nutrizionali nell'ambito della 'Dieta mediterranea', Corso di formazione 1bis dgr 347/11 - PSR CAMPANIA 2007/13 - Mis. 111 "La relazione tra alimentazione e salute umana: il caso della cucina tradizionale delle aree interne e il suo riferimento alla dieta mediterranea", [Faicchio (BN), 14 maggio 2014], http://aspa.unitus.it/matassino/1_elenco_pubblicazioni_Matassino.pdf.

71) G. Maiani et al., Risorse Alimentari e Dieta Mediterranea. , in: Atti IX Convegno Nazionale sulla Biodiversità . Biodiversità [eds. G. Calabrese et al., Valenzano, Bari, 6-7 settembre 2012], vol. II "Biodiversità, Alimenti e salute", 3 (CIHEAM-IAMB, 2013).

72) J. Salobir, Krmiva, 49 (3), 147 (2007).

73) V. A. Sironi, Il cervello ci indica il cibo E il cibo modifica il cervello, Avvenire, 2 agosto 2014, 3.

74) E. Sereni, Storia del paesaggio agrario italiano. (Laterza), (Bari, 1961).

75) C. Nardone, personal communication, (2013).

76) M. White et al., Journal of Environmental Psychology, 30 (20), 482 (2010).

77) D. Matassino, Bioterritori intelligenti, Workshop "L'officina dell'ingegno", [Petrella Tifernina (CB), 24 luglio 2013], http://aspa.altervista.org/ ; link : archivio Prof. Donato Matassino.

78) G. Sabato, Mente e cervello, 103, 36 (2013).

79) P. Bernardi, Nel nome della luce, I luoghi dell'infinito, 182, 8, (2014).

80) F. Cardini , Rosso, giallo, blu: simboli in festa, I luoghi dell'infinito, 182, 14 (2014).

81) P. Daverio, Visioni cromatiche , come un'avventura, I luoghi dell'infinito, 182, 5 (2014).

82) Z. Bauman, Vita liquida, Laterza, Bari- (Roma, 2006).

83) B. Forte, Lo stupore chiamato per nome, I luoghi dell'infinito, 181, (2014).

84) M. G. Riva, Cosí la via della pulchritudinis conduce al Mistero, I luoghi dell'infinito, 181, editoriale (2014).

85) D. Matassino, Filosofia strategica gestionale di un bioterritorio allevante il "bovino Grigio autoctono italiano" (già "Podolica"), in: Atti Convegno "Dal pascolo alla tavola: sicurezza e qualità dei prodotti 'podolici' [Zungoli (AV), 29 ottobre 2011], (DELTA 3 Edizioni), 31 [Grottaminarda (AV), 2013].

86) T.M. Vogel et al., Nature Rev. Microbiol., 7, 252 (2009).

87) A. Bose et al., Nature communication, 5, 3391 (2014).

88) D. Matassino, La biotecnologia tra libertà di ricerca e regolamentazione, in: Atti Secondo Convegno Regionale di Bioetica per la Scuola "Prolungamento della vita umana e ingegneria genetica" (ed. M.A. La Torre 'Bioetica e diritti umani', Caserta, 27 marzo 2004, (Luciano Editore), 143 (2005).

89) D. Matassino, Laicità della scienza, in: Atti Ciclo seminariale "Fede e Ragione", (Eds G. Di Palma e P. Giustiniani 'Teologia e Modernità - Percorsi tra ragione e fede', Napoli, 11 febbraio 2008), (Pontificia Facoltà Teologica dell'Italia Meridionale), 127, (Napoli, 2010).

90) P. Mantegazza, Elementi d'igiene, (Casa editrice Gaetano Brigola), (Milano, 1867).

Printed in the United States
By Bookmasters